Slings and Arrows

Mark Denny

Slings and Arrows

The Science of Ancient Projectile Weapons

Springer

Mark Denny
Victoria, BC, Canada

ISBN 978-3-032-08562-7 ISBN 978-3-032-08563-4 (eBook)
https://doi.org/10.1007/978-3-032-08563-4

This Springer imprint is published by the registered company Springer Nature Switzerland AG
The registered company address is: Gewerbestrasse 11, 6330 Cham, Switzerland

For Dr. Robert O'Connor, Dr. Markus Sikkel, Dr. Michael Metcalfe, Dr. Saibishkumar Elantholi Parameswaran and their people, without whom this book would not have been written.

Preface

I didn't know it at the time, but I suppose that the seed of this book was planted over 20 years ago, with a technical article on counterweight trebuchets published in the *European Journal of Physics*. These giant siege engines were, it seems, amenable to physics analysis—that is, the internal workings of the machine could be analyzed and their action predicted. Given the beam dimensions and weights, counterweight, projectile weight, and sling length I could tell you the launch speed of the projectile (typically these monsters fired a 200-lb rounded rock), sent silently skyward. The trajectory of a trebuchet stone might have terminated not so silently against the curtain wall of a castle. It was the internal workings of the machine that grabbed me—my work was done just as soon as the rock was airborne. Yet, well, the fate of the stone—and of the castle and its inhabitants—was hard to ignore.

Years later Mike Fulton, a professional historian and archaeologist who is an expert on medieval warfare, reached out for information on trebuchet capabilities, and so that early *European Journal of Physics* article provided him with answers. Meanwhile I had written other papers on a smorgasbord of physics topics (steam engine governors, pendulum clock escapements, onagers, longbows, curling rocks, waterwheels, antilock brakes, bat echolocation, albatross flight, gas gun dynamics, non-returning boomerangs, funicular bridges, sling dynamics, earthquakes, atlatls, slings...). You may have noticed, in amongst that catalog of arcane oddballs, other ancient weapons. You see, not all of them had been previously analyzed as to their internal workings, and none to my satisfaction. Thus, a survey of the physics literature would have revealed very technical accounts of how longbows worked, but these invoked beam theory and finite element analysis. You could have read detailed analyses of boomerang flights that required full-on computational fluid dynamics, airfoil theory and more finite element analysis. Nothing wrong with these works—they gave the right answers—but were inaccessible and shed little light. My approach has been to provide much simpler calculations that are approximate, and insightful.

It had been a little frustrating for me to fit the physical analyses of ancient machines into the format required by physics journals: be brief, concise, mathematically unambiguous and so, unavoidably, be very dry. I had wanted to include history and technological development which to my mind set the scene and rounded the accounts nicely but to control costs the physics journals—not unreasonably—require brevity and strict relevance. A published physics paper is like a zipped file—the reader is expected to have the mathematical acumen and physics background to unzip the content him/herself. History cannot be zipped, only curtailed. Several times I had to rewrite papers because they were too long, embellished with 'extraneous' material. I wanted to weave the science and the history together, to tell a more complete story of these ancient machines.

Last year I asked Springer if they would be interested in a monograph on the subject of ancient weapons and they said yes. So here we are.

There are a surprising number of Mike Fultons out there. Some are like him—professional historians or archaeologists who want to know performance capabilities. Others are hobbyists who inhabit many countries and Youtube, who are passionately interested in bows or boomerangs or spear throwers or slings or siege engines. Yet others are fellow physicists or science and technology buffs or amateur historians who are just curious (perhaps in both senses of the word). They are the software engineers or accountants who yearn to be Parthian horse archers (they will like Chap. 4) or Han crossbowmen (Chap. 5) or Balearic slingers (Chap. 2). They are happy to share their passion online: there is a German society for boomerang throwers [5] and there are American Youtubers enthusing about, and designing and throwing, *rabbitsticks* (the version of boomerang that interests us, and the subject of Chap. 3) [6]. There are atlatlists (an odd word that atlatl throwers—of which there are many across the world—have coined to describe themselves) who stage competitions [7]. Atlatls are the main subject of Chap. 1. There are engineers who construct ballistas (Chap. 6) and counterweight siege engines (discussed in Chap. 7) in England, France, Germany and Denmark, for the benefit of tourists and historical reenactments [8]. This book is for all these splendid folk—who themselves form the subject of the final chapter, about the role and popularity of our old weapons in the modern world.

The plan is to explain the inner workings of ancient projectile weapons. Here 'ancient' means from the Paleolithic to the rise of gunpowder armaments; 'projectile' means ballistic ordnance—aimed projectiles which flew through the air to strike a target some way off; 'weapons' means tools for hunting or warfare. Some readers may have noted a preference herein for the internal aspects of these weapons, and indeed I will be concentrating upon the machinery—their construction and workings. Their purpose was to send a projectile of some sort, as fast as possible, in a chosen direction. My main goal is to show how this was achieved. Of course you would also like to know what happened afterwards—what was the target range, and what was the effect of the projectile upon it. Range calculations are made in the first of several Technical Appendices. Range estimation is common to all our projectile weapons, and so we deal with it at the start.

The interesting and sometimes surprising history of these machines—their cultural context, usage, military effectiveness and influence—has been provided to me by several experts, some of whom are historians and archaeologists and others hobbyists and history enthusiasts. Beginning with the professionals, thanks to Prof. Duncan Campbell of Glasgow University, Scotland, for answering questions about onagers and ballistas, and providing me with references. Professor Michael Fulton of the University of Western Ontario, Canada, patiently answered many, many questions on medieval siege engines—thanks, Mike. I am grateful to Prof. David Graff of Kansas State University who has provided all kinds of technical details about Chinese crossbows, patiently answering my many questions.

Thanks to Dr. John Hayward, an archaeologist at Flinders University in Australia who has provided detailed answers to my questions about boomerangs, along with references to technical articles. Also in Australia and concerning boomerangs, Dr. Philip Jones of the South Australian Museum fully replied to my questions and urged me to take up boomerang throwing. I must thank Prof. Clifford Rogers of the United States Military Academy at West Point for useful and out-of-the-way information on archery in the Middle Ages in general, and the English longbow in particular. Professor John Whittaker, an anthropologist at Grinnell College, Iowa, has sent me many articles about atlatls, encouraged me to take up atlatl throwing, and answered a wide variety of questions about these spear-throwers—thank you John.

The very knowledgeable hobbyists and history enthusiasts who have significantly helped my research begin with Luca Bottazzi, who has been very patient with me, answering lots of questions about the Japanese longbow and Japanese medieval armor. He maintains the excellent website https://gunsenmilitaryhistory.wordpress.com/. I am grateful to Lewis Everett of Archaic Arms, an expert slinger who answered my many practical questions about sling use. Professor Chris Harrison is a professional computer scientist but also an expert on the sling, and I thank him for answering my questions about slings and their place in military history. I asked 戰國春秋 of http://greatmingmilitary.blogspot.com lots of questions on ancient and medieval Chinese military history, and learned a lot—many thanks.

I am grateful to Sam Harrison at Springer, who guided me through the book proposal/acquisition phase, and to Sridevi Purushothaman who coordinated everything to do with publishing. You made it easy for me, guys.

My friend of 50 years from graduate school days, David Derbes, kindly sent me out-of-the-way references when I told him the subject of my new book. Last but definitely not least, my wife Jane has been a book-widow for about a year, for the fifteenth time. Thank you both.

Victoria, Canada Mark Denny

Competing Interests The author has no competing interests to declare that are relevant to the content of this manuscript.

Introduction

The *FGM-148 Javelin* is a man-portable anti-tank weapons system. The javelin is a thrown stick—the earliest were hafted with a stone tip. Trick question: which of these two projectile weapons was the first to be introduced to the world by Homo sapiens? Answer: the former, in 1996. The javelin had been introduced some half million years earlier [9, 10, 11], most likely by Homo heidelbergensis, but certainly not by our species.[1]

Neither of these weapons will be covered in this book, which will look at pre-gunpowder human weapons not necessarily in chronological order but instead in order of increasing complexity or technological sophistication. None of our weapons are as old as the javelin, but some of them originated in the Stone Ages [1, 2, 3, 4]. Figure 1 is a rough timeline for the main weapons featured in this book (the *FGM-148 Javelin* would be at 'Now'.) My aim is to convey what I have learned about ancient weapons by analyzing the dynamics of their behavior—in particular, how they are launched—and placing this analysis in historical context. The two subjects—physics and history—do not overlap much in the literature, but here they must. (To take a simple example, the physics of firing a bow depends on the materials available at the time and upon the technology of manufacturing bows. The fiberglass and carbon fiber of modern compound bows were unavailable a thousand years ago, when bows were the main projectile weapon of an army; instead they used wood, or a combination of wood, horn and sinew.) Examining this overlap sheds

[1] Anatomically modern humans arose about 300,000 years ago.

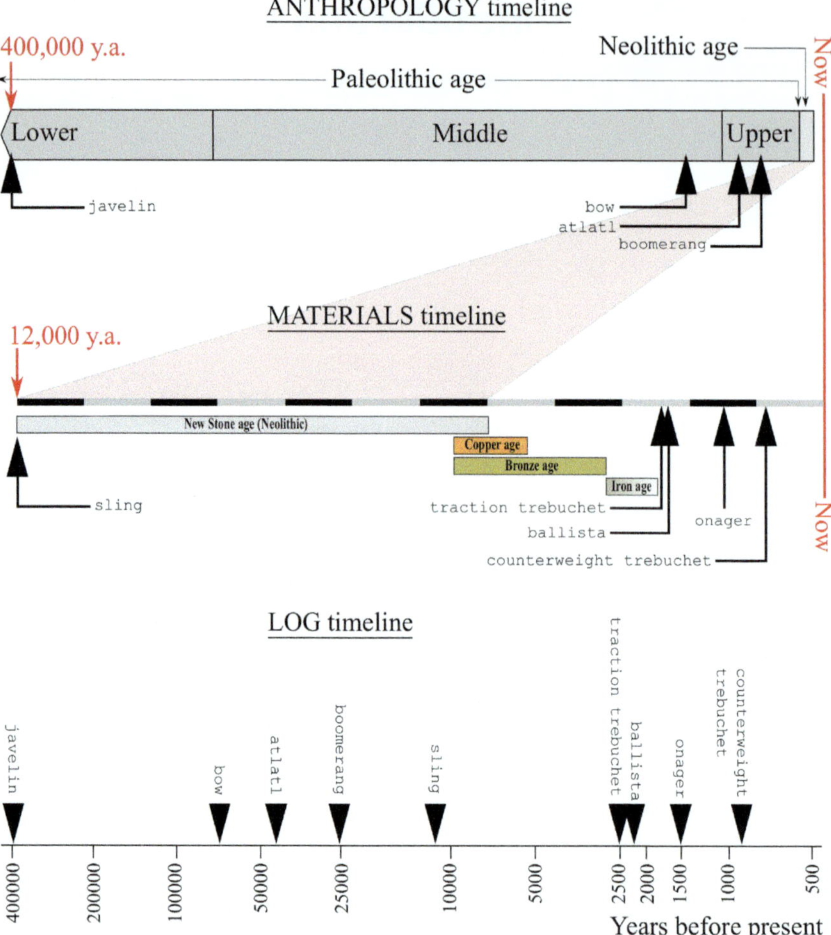

Fig. 1 Three timelines: anthropological timeline (four of our weapons were first developed in the Paleolithic age), materials timeline (the remaining weapons generally required iron), and a logarithmic timeline to show the age of all our weapons on a single line. These are age estimates for widespread use of our weapons and are approximate—take them with a pinch of salt. (Thus for example the onager is listed as originating in the middle of the fifth century CE, though it was certainly known to the Greeks several hundred years before.)

light on these weapons, and for me has increased admiration for the many people who developed them and for the myriads who have learned the skills to use them.

Historians and archaeologists both appreciate physical analyses because they read about or dig up ancient weapons and would like to know how well they worked: the range of a sling, the reliability with which a siege engine could hit the same spot on a castle wall, the rate of fire of a crossbow, the performance difference between short bows, longbows, composite bows, recurve bows, etc. An archaeologist may dig up the remains of a Balearic sling and, having read the classical authors about the performance of these weapons, wonder if the claims made were exaggerations, or true—could a slinger from two thousand years ago really hurl a projectile over a quarter mile? The answer (as you will see) is yes. Certain slings can, in the capable hands of an expert, generate a high enough launch speed to throw a lead sling-bullet (as sling projectiles are called) over 400 m. (They are not the only weapon that appears in the pages to follow that can launch a projectile such a distance.)

This is a good place to note that I will use metric units in the Technical Appendices (TAs, henceforth) where the math is set down, and both English and metric units in the main text, with conversions given as needed—400 m is about a quarter mile. The reason for including the quaint English units in the main text is simply because they are still widely used in the United States generally, and amongst hobbyists around the world (thus, archers and archery websites still measure arrow weight in grains and bow draw weight in pounds). Full disclosure: in Chap. 6 I will have occasion to use Ancient Greek units too.

Ah, yes—the math. To present the very interesting history and development of ancient weapons, and also to present an insightful analysis of their dynamics, we need to fit the square peg of physics into the round hole of history. I have written fourteen popular science books, each with mathematical equations within, and it took me five or six books to learn that the best way to combine history and physics in a readable way is to relegate the hard-core math analysis to appendices, and to present in the main text only the conclusions that have been drawn from the math. If you, like me, are something of a geek who craves the fundamental principles then the TAs will provide enough mathematical detail to derive most of the claims made in the main text.[2] On the other hand your author is acutely aware that some readers would rather endure root canal work than wade through calculus. If you are one of these readers for whom a summary will suffice then you may omit the TAs entirely; the only price to pay for such a literary appendectomy is that you will have to take my analysis on trust. The main text will include only High School physics ideas, and no explicit calculations. The TAs contain undergraduate-level physics and math.

[2] To fully reconstruct the physical analysis you may need to refer to the technical literature referenced in the Appendices. These are mostly papers I have written or primary sources upon which my stuff was based.

The history of these venerable devices in the pages to follow emphasizes technological evolution of ancient projectile weapons, whereas the math analysis emphasizes the projectiles' internal and external trajectories. It is conventional in the literature to divide a ballistic trajectory into three parts: internal ballistics deals with the internal movement up until the projectile parts from its launcher. External ballistics looks at how it flies through the air, and terminal ballistics examines what happens when it strikes its target. For gunpowder weapons I have in the past labeled these three parts the bang, whizz and thud phases [12]. By concentrating on the launch phase for our ancient weapons I can convey the techniques and skills that the slinger/thrower/archer/siege engineer must develop in order to strike a target at distance. There are exceptions: we will need to consider external ballistics to understand the boomerang, for example, but otherwise our discussions of the whizz and thud phases will be limited to range estimation. (Limited and squirreled away: the range of a projectile, given its initial velocity, is the subject of a preliminary appendix, TA-0.)

Finally I should say something about the processes employed by the soldiers or engineers or hunters of old to develop their weapons. They lived their lives at least half a millennium before the scientific revolution, which kick-started the industrial revolution and the modern research and development model of human progress. Their ideas and experiments and thought processes were carried out individually or as part of a small group. They had no theoretical understanding—no Newtonian mechanics or modern aerodynamics—to guide them. Instead they worked empirically. A good example of the empirical approach is the alleged 'five-minute rule' adopted by those master-builders, the Romans: if a building stands for five minutes, then the design is good. The epitome of empirical development in this book is the non-returning boomerang, as we will see. The thrower[3] had to tune their boomerang, by whittling it or twisting it or polishing it, so that it would fly just right. More generally our thrower of old learns how best to launch their projectile, and then they might adjust some aspect of their throwing action, or make a slight tweak in the projectile shape or size. This would result in a change in the trajectory for better or worse. They would learn from their results and adjust further to improve performance. This is empirical progression: it requires no theoretical knowledge but instead is based upon a great deal of experience and testing and intelligent cogitation.

It is important to note that empiricism is not at odds with the scientific method, indeed it is science. The thrower performs experiments with their

[3] Henceforth for brevity we will refer generically to the person who launches our missiles—boomerang, sling-bullet, javelin, arrow, bolt, or rock—as the *thrower*.

weapons and interprets the results. They tinker with the weapon parameters (lengths, projectile weights) and find that their equipment performs worse for the same throwing effort, and so they throw away the modification. They tinker again, differently, and this time find an improved performance for the same effort—they keep this modification. Such an approach to innovation or design evolution is empirical science. A very simple example (we will encounter many more in the pages to follow): you throw a rock and find that you can achieve a certain distance—say 20 yards. You throw it again with the same effort but at a different elevation angle, and the rock flies a distance of 23 yards. We know that this means the second throw was directed at an elevation closer to 45° than the first throw, but Ug and Ugga from the Paleolithic didn't know about degrees—however I will bet that they knew how to throw a rock.

And they passed on their accumulated knowledge to their successors. Isaac Newton once said: "If I have seen further, it is by standing on the shoulders of giants." [13] Our ancient technologists built on the knowledge of their predecessors, accumulated over centuries or millennia, added their bit and passed it on, in exactly the manner that Newton meant. This book contains my accumulated knowledge of the science of ancient projectile weapons, and I pass it onto you—I hope you find it entertaining and instructive, and perhaps useful.

References

1. L. Metz, J.E. Lewis and L. Slimak, "Bow-and-arrow technology of the first modern humans in Europe 54,000 years ago at Mandrin, France," Sci. Adv. 9 eadd4675 (2023).
2. J.E. McClellan and H. Dorn, Science and technology in world history: an introduction (Johns Hopkins University Press, Baltimore MD, 2006) p. 11. ISBN 978-0-8018-8360-6.
3. M. Lombard, "Re-considering the origins of Old World spearthrower-and-dart hunting," Quat. Sci. Rev. 293 107677 (2022).
4. P. Valde-Nowak, A. Nadachowski and M. Wolsan,"Upper Palaeolithic boomerang made of a mammoth tusk in south Poland," Nature 329 436–438 (1987).
5. The German society is Deutscher Bumerang Club e. V. There is also the Societa Italiana Boomerang Sportivo, the Australian Boomerang Association, the Associacao Brasileira de Bumerangue, the Dansk Boomerang Forbund, the British Boomerang Society, etc. For a more complete list see the webpage of the United

States Boomerang Association at https://usba.org/international-boomerang-associations/.

6. See for example, the Youtube videos Rabbitstick! and The Rabbitstick: The Not So Primitive Hunting Weapon.
7. There are many Youtube videos about atlatl throwing, such as Atlatl Demonstration and The Atlatl: Most Underrated Stone Age Tool?. See the Videography at the end of the book.
8. An impressive full-scale reconstruction of a trebuchet is shown launching missiles on the Youtube video Unleashing a Medieval Trebuchet on a Wooden Palisade.
9. H Thieme “Lower Paleolithic hunting spears from Germany,” Nature 385 807-810 (1997).
10. B Bower 2012 “Oldest examples of hunting weapon uncovered in South Africa,” Science News 11/15/12.
11. J Wilkins et al 2012 “Evidence for early hafted hunting technology,” Science 338 p 942.
12. M Denny 2011 Their Arrows Will Darken the Sun (Baltimore: Johns Hopkins University Press).
13. Isaac Newton letter to Robert Hooke, 1675. Available online at https://discover.hsp.org/Record/dc-9792/Description#tabnav. Newton may have been having a dig at his competitor Hooke, here (Robert Hooke was of diminutive stature.)

Contents

Acronyms

BCE	(Years) Before Common Era
CE	Common Era
CM	Center of Mass
CW	Counterweight
FCWT	Fixed-Counterweight Trebuchet
NRB	Non-Returning Boomerang
RB	Returning Boomerang
SCWT	Swinging-Counterweight Trebuchet
TA	Technical Appendix

1

Spear-Throwers: The Ancient Atlatl

Chapter Summary The origins and distribution of atlatls, and their historical persistence, are discussed. The simplest physical principles are presented: the atlatl acts like a lever, so the launch speed of its dart projectile exceeds that of the same dart thrown without the aid of the atlatl. Refinements (such as thrower body motion, and flexible dart shaft) require more subtle physics. The nature of approximations in physics is outlined and the case made for their usefulness for all of the ancient projectile weapons examined in this book. The significance of atlatls in Mesoamerica at the time of the Conquest is summarized.

"Atlatls are ancient weapons that preceded the bow and arrow in most parts of the world and are one of humankind's first mechanical inventions. The word atlatl (pronounced AT-latl) comes from the Nahuatl language of the Aztec, who were still using them when encountered by the Spanish in the 1500 s."—(World Atlatl Association) [1].

1 Stone-Age Surface-to-Surface Missiles

Of the six classes of ancient projectile weapons that we will investigate in this book, the first three are—let's be blunt—sticks and string, and nothing more. Little else was available back in the day, in terms of material with which to make a munition of any kind. We begin in the late Paleolithic and for the next several thousand years (i.e., the first three chapters) all we have to work with is wood, stone, and string.

There was for sure a weapons industry, right from the get-go: those javelins bequeathed to us by other hominids (or, perhaps more likely, independently invented by us) were, no doubt, carefully chosen sticks or tree branches which

M. Denny, *Slings and Arrows*,
https://doi.org/10.1007/978-3-032-08563-4_1

would have then been modified from the natural state: bark stripped, points added, heated for hardness, etc. We will not consider javelins further, in isolation. We begin here with the spear-thrower, a machine—a lever—for increasing javelin range, speed, and penetrating power. Given that range was wanted, it makes sense that the javelins used as spear-thrower projectiles would be lighter than hand-thrown javelins and much lighter than spears. From now on we call these projectiles *darts.*

Darts were lighter than other javelins but they had a greater launch speed, as we will see, and so their momentum will have been at least as large. Large enough to kill deer, bison, or even mammoths from a range of 10–15 yards, according to recent experimental and archaeological research [2, 3]. The incentive for developing a weapon that could kill a large animal (or a human) from distance is obvious: the hunter or warrior would have wished to minimize risk to himself. Or herself. To realize the projectile weapon—to get from wishing for some method of feeding your family without being gored or trampled in the process—took intelligence. Using a tool such as a hand scraper or a hook was one thing (other primates and some birds can do that) but to make the leap to *constructing* a tool that did the job is quite another. Tool construction (in this case, modifying a piece of wood, and not just picking it up off the ground) takes that uniquely hominid combination of capabilities: brains for planning and hands for realizing the plan—to make the tool and use it.

2 Origins

The atlatl is an ancient tool [4–6]. Our earliest definitive archaeological evidence, at the time of writing, for spear-throwers is about 17,500 years old: a reindeer antler from a cave in the Dordogne region of France, fashioned to form a spear-thrower. That form is simple to describe: a short stick with a handle at one end and a shaft or spur at the other, the purpose of which is to hold the butt end of a dart while it is being thrown. A much younger example (the form hardly changed over millennia) is shown in Fig. 1.

The spur is on the right end of the high-status atlatl shown in Fig. 1. The larger protrusion on the left holds the dart clear of the atlatl shaft, which makes the throwing action easier. This protrusion is not essential and many atlatls, especially early ones, lacked it. (Note also from Fig. 1 that in the New World atlatl use lasted beyond the Stone Age.)

The fact of the oldest remains being found in France does not mean that the atlatl was first developed there: its use spread across the world over the millennia of prehistory, from wherever a smart hunter first thought of it—or perhaps

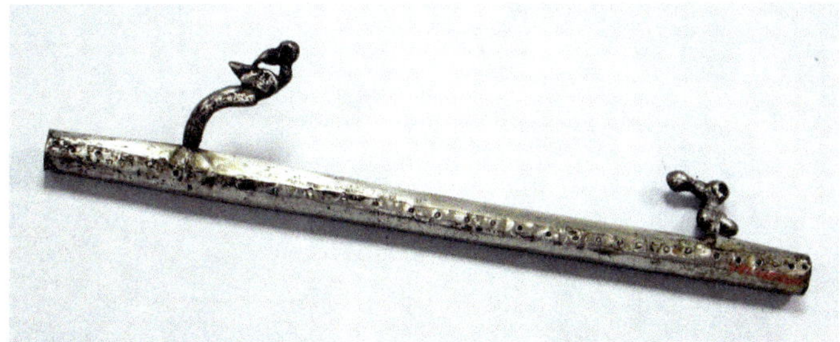

Fig. 1 A hammered and cast silver atlatl from the Chimù culture of Peru, twelfth to fifteenth centuries. L = 10 inches. Image courtesy of the Metropolitan Museum of Art (NY) Open Access policy

it developed independently in many parts of the world at different times. Thus, for example, there is indirect evidence from 42,000-year-old skeletons in Australia. These show that the owners (I guess that the person who is built up around a skeleton can be called its owner) suffered from what we would think of today as "tennis elbow." It seems unlikely that Paleolithic Australians played tennis, but we know that repeated throwing of atlatls leads to very similar arthritic elbow injuries [7].

We can get an idea of how far the spear-thrower spread—in space and time—from the names given to it by different people. We are going with the Nahuatl word and calling this weapon *atlatl* in this book, but in the Dhurug language of southeastern Australia it is called *woomera.* The woomera is broader than a simple stick and is, like the boomerang, a multi-purpose tool for native Australians. The Spanish conquistadors found the weapon used widely among people native to the Andes and called it *estólica.* It was often decorated with colorful feathers. Indeed, throughout the world spear-throwers were in many cases carved and decorated. The ancient Greeks had a similar weapon which they called *amentum*; the stick was replaced by a leather throwing strap, which served the same purpose (though not so well) of extending the reach of the thrower. The Tlingit people of Alaska called it *she-àan* meaning "sitting on a branch." Chinese from the Song Dynasty (960–1279 CE) called their unusual spear-thrower *bian jian* meaning "whip arrow." It was a very long (10 ft) bamboo staff and required two operators.[1]

You have likely realized—correctly—that the forms taken by atlatls around the world were as varied as the names given to them. Here is another, rather

[1] *Woomera* is the name of a major Australian aerospace facility—a nod to the ancient missile-launcher. The *bian jian* was a hybrid of atlatl and staff-sling (Chap. 8). The lower end of the staff was swung by one operator; the other end was attached to a cord—so the whole thing resembled a whip. The dart was initially attached to the far end of the whip, by the second operator.

Fig. 2 A "butterfly" bannerstone made of slate, made in eastern North America sometime between 6000 and 1000BCE. The two notches are connected by a hole through which the atlatl passed

mysterious variation: in North America atlatls often had a small *bannerstone* attached to the middle of the shaft. The practice of attaching bannerstones dates back into prehistory (see Fig. 2). They are mysterious because their function is unknown, yet clearly (from the effort invested in making them, and the fact that bannerstone mass is a significant fraction of atlatl mass) was deemed necessary. There are different claims by experts as to their purpose[2]: perhaps cultural, with no physical function—maybe it brought luck to the thrower—or perhaps the bannerstone inertia steadied the throw, increasing accuracy or improving dart stability. Given the intricate carving and polishing of many of these stones, superfluous to any mechanical purpose, I will go with a cultural function [8]. Bannerstones are, I think, unique to North America, but other cultures considered it necessary to adorn their atlatls with decorations, typically feathers. Perhaps such adornments attest to the importance attached to atlatls as useful—essential—tools.

Such spatial variations in the form of a tool that was widely used around the world for a very long time suggest that the developments occurred in isolation independently of one another, but also points to the ingenuity of early humans and to their strong motivations to develop a throwing weapon.

Whatever the form taken, the atlatl was for thousands of years a very effective hunting implement and in the New World (particularly Central and South America) a projectile weapon of war. Thus, in seventeenth-century Brazil we have evidence from European writers as to their use by native people: *"The atlatl, as used by these Tarairiu warriors...[threw a dart] made of a two-meter*

[2] I mean the purpose of the bannerstones, not the purpose of the experts. But you knew that.

long wooden cane with a stone or long and serrated hard-wood point, sometimes tipped with poison. Equipped with their uniquely grooved atlatl, they could hurl their long darts from a great distance with accuracy, speed, and such deadly force that these easily pierced through the protective armor of the Portuguese or any other enemy." (Prins, 2010 [9]). See Fig. 3.

A common assumption among archaeologists and anthropologists is that projectile weapons evolved most naturally in the following manner: first came the spear, then the dart, and then the arrow [10, 11]. That is to say, a people with spears threw them, then adapted their thrown spears so they flew better—further or more accurately—thus generating darts. From atlatls and darts it is a

Fig. 3 "Portrait of a Tapuya male with atlatl hunting gear," Albert Eckhout 1641

short step to bows and arrows or at least a shorter step than from stabbing spears to bows. This assumption seems very reasonable to me, although it has been pointed out that such an evolution has not been universal [12]. In some parts of the world (Africa, for example) there is no evidence of atlatls but plenty of bows. There is some slight evidence for the exclusion of one in certain societies where the other dominated.[3] Such variations on the theme are to be expected for a widespread and popular piece of technology.

It is probable that atlatls were used most widely for hunting and not for organized warfare. Aztecs made use of atlatl throwers in the initial stages of battles, much as the Roman armies used slingers. Elsewhere the "common assumption" likely had a corollary: bows and arrows supplanted atlatls and darts.

3 How They Work

The atlatl throwing action is more complicated than that for a spear, and takes some practice to master.

3.1 Basics: It's a Lever

It's not difficult to estimate the launch speed of an atlatl dart, given reasonable estimates of dart weights and atlatl dimensions, big game kill energies, and Major League Baseball fastball pitch speeds. Ok, that sounds odd, so let me unpack it all. I will go through the method for this first calculation in some detail (the math, as promised, is all in the appendix TA-1 and ref. [14]) so that you see how it all works. We can get a lot of what we want from a few simple steps, if we are happy to settle for approximate answers.

The action of an expert thrower (*atlatlist* appears to be the word favored in the literature) launching a dart with the aid of an atlatl is shown in Fig. 4. Note that their action applies the force of the throw along a straight line, to a good approximation. It is easy to see why this happens: any variation of the force in different directions during the throw—say to the left and to the right—will reduce the energy that is transmitted to the dart for forward flight. The thrower/atlatlist will work out the best way to launch the dart, to attain maximum launch speed for a given effort, and this way will be such that the

[3] In Australia bows were little used, though northern Australians certainly knew of them. The atlatl and the boomerang were the preferred long-distance weapons on this continent. There is a suggestion in the literature that the mechanical properties of Australian flora were such as to make them inferior materials for constructing bows [13].

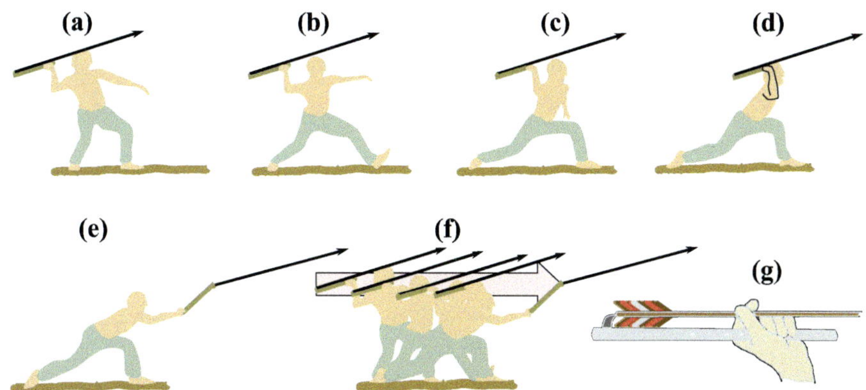

Fig. 4 Expert "atlatlist" action. **a**–**e** Sequence of thrower's position, from beginning to release of the dart. **f** The images superimposed—the force exerted by the thrower is acting approximately along a straight line. Such an action saves energy. **g** Close-up showing the spur holding the dart butt end, and the thrower's grip, with a finger or thumb holding the dart in place during phases (**a**)–(**d**) of the throw

force is exerted along a straight line. From this linear force we can estimate the energy imparted to the dart. In TA-1 you can see why the launch speed of the dart exceeds that of the same dart thrown without the aid of an atlatl by this amount:

$$v_{\text{atlatl}} \approx v_{\text{thrown}} \sqrt{1 + \frac{2l}{d}}. \tag{1}$$

Here $\approx$ means "is approximately equal to" (you will see this symbol a lot in the pages to follow), l is the length of the atlatl. In Eq. (1) d is the internal distance of the throw without the atlatl—*not* the range of the dart after release, but the distance covered by the accelerating hand holding the dart before it is released. That is, imagine the thrower of Fig. 4 is holding the dart like a javelin, without the atlatl: they will move their throwing hand a distance d from steps (a) to (e), from the beginning of the throw to the instant of release. With the atlatl you can see from Fig. 4 that this distance is increased by approximately twice the atlatl length. The atlatl is being used as a kind of lever, but not in the sense of generating mechanical advantage, but rather by extending the reach of the thrower.

To complete the calculation, we need to know the parameter values of Eq. (1): a reasonable choice for atlatl length is $l = 0.5$ m, say 20 inches. A reasonable choice for the distance over which the throwing hand is accelerated prior to release is $d \approx 1$ m. What about the speed v_{thrown} in Eq. (1)? This is where the fastball pitcher comes in. Consulting TA-1 you will find that a 100 mph fastball thrown by a top-rated MLB pitcher has about twice the energy that a mere mortal can impart to it. Assuming that atlatl throwers in the past were mere mortals, therefore, we can estimate v_{thrown} and so, from Eq. (1), find that $v_{\text{atlatl}} \approx 25$ m/s—call it 55 mph. What damage could a 1-lb dart, moving at this speed, do to a target? I do not have an ancient atlatl to test for myself, nor the expertise to use it effectively if I did have one, and expert atlatl throwers are thin on the ground in my part of the world (but see Chap. 8). So I have trawled through many online sources and it seems to me that the nearest equivalent would be modern Internet-savvy bowhunters. They have data posted online that fits the bill—telling how much energy an arrow needs to kill a deer (25 ft-lb) or a grizzly bear (65 ft-lb or more).

Hunters in North America seem to go in for the quaint, anachronistic, and ridiculous energy units used centuries ago in England (foot-pounds or ft-lb). Converting to metric units reveals that an atlatl could kill an elk, or antelope. It would seriously wound a grizzly—two atlatl darts would dispatch one. A team (pack?) of Stone-Age hunters could fell a mammoth.[4] This method of estimating kill capability (energy of the projectile) is rough and ready. In later chapters, we will see that a different metric—projectile momentum—is more useful, at least if the targets are armored (say Persian cataphracts—heavy cavalryman—rather than moose).

A number of assumptions have been quietly inserted into the calculation just summarized—and outlined in more detail in TA-1—but all are reasonable and lead to plausible predictions for the capability of atlatl throwers of days gone by. A couple of them hunting a bison could kill it from 10 or 15 yards away. One of the throwers could kill the other from twice or three times that range, if they were accurate enough. I am sure it's been done, many times.

3.2 Subtlety: It's More Than a Lever

Equation (1) quantifies the benefit to be gained by using a spear-thrower: dart launch speed is increased when compared with throwing the same dart using an arm alone. The longer the atlatl, the more that launch speed increases. (Up to a point: too long and the atlatl will flex, making the throw more difficult.) How can we be confident that the approximations I have made in deriving Eq. (1) are realistic, so that the solution truly represents an atlatl throw? How do we know that in making approximations I am not throwing out the baby with the bathwater? This is an important question that arises for most of our ancient projectile weapons, so it is perhaps worthwhile to deal with it here in the first chapter.

---

[4] The key seems to be striking the prey animal where there is no bone so the dart can penetrate deeply. J Whittaker, personal communication July 2025.

The siege engines of Chaps. 6 and 7 are much more complex machines than the atlatl, or the sling of Chap. 2 (the "sticks" referred to earlier). Yet ironically their physical analysis is easier and less approximate. The reason is that siege engines are made of thick pieces of wood, and metal, which deform only a very small amount during the launch of their projectiles. I can thus make the eminently reasonable approximation for them that the component parts are rigid. To an investigating physicist this is joyful news because it immediately reduces the complexity of the system under investigation by reducing the number of degrees of freedom (the number of variables). For our simple sticks, however, the full system under investigation is not just the atlatl or sling, it is the atlatl plus atlatlist, or the sling plus slinger. The human body is the (very flexible and thus difficult to quantify) source of power that drives these weapons. This flexibility, and consequential difficulty of analysis (or even of posing the question), is the reason why many and important approximations need to be made.

So, again, how can we be confident that the approximations that I (and others) have made lead to realistic solutions? Sometimes it is obvious that an approximation is not going to lead us astray: saying that a thick wooden beam that forms part of a trebuchet is rigid, for example. Strictly speaking it is not—it may bend or twist or compress by a millimeter, but if that is only a thousandth of the beam length then any error that is a consequence of our rigidity assumption is negligible. Likewise with the implicit assumption I made that the atlatl is rigid, or that the butt end of the dart doesn't slip during the release. (It *could* slip and so fly off prematurely, but if so then the resulting trajectory would be obviously due to slip. The dart cannot slip so as to affect the launch speed significantly without us knowing that it has slipped.)

Another obviously reasonable assumption that was implied in TA-1, leading to Eq. (1), is that there is no friction acting between dart butt and atlatl spur, and no aerodynamic drag that slows down the motion of dart thrower, atlatl and dart during this internal ballistic phase of dart motion. Of course there is air resistance but its effects are small enough to ignore (this is patently *not* true for external ballistics, as we see in TA-0). Knowing when and which approximations to make in physics is part of the art of the subject.

The main approximation that I made in the analysis outlined in TA-1 that is not obviously right (indeed, for some darts it is obviously wrong) is that the dart is rigid, that it remains straight throughout the launch phase. Some atlatl darts are a couple of meters long, or more, and thin enough to bend under gravity, so when stressed by the force of the throw they buckle [14, 15]—there is video evidence aplenty for this well-known phenomenon.[5] So how do I get away with the gross oversimplification that darts are rigid?

This is a different type of approximation. I am not saying "the approximation is small and so the errors it introduces are small" as was the case for trebuchet beam rigidity, for example. Here I am saying "the approximation may be significant but it does not generate significant errors in the center-of-mass (CM) motion." In other words, the dart may flex back and forth but this flexing does not change the location of its CM and so predictions about CM motion (such as dart speed) are not subject to significant error. For example, high-frame-rate movies of a golf ball being struck show that the ball most certainly does not remain spherical when slammed by the club head—it flattens like a pancake—but this brief distortion of shape does not influence in any significant way the predictions we make about subsequent golf ball motion.

That is quite enough discussion about approximations. Returning to atlatl darts, there are a couple of interesting consequences of dart buckling [14]. First, buckling may (counterintuitively) improve the thrower's aim, rather than make it worse. Second, dart vibrations during flight (caused by buckling during launch) reduce the dart maximum range somewhat, but not by much—only a little of the dart energy is diverted to the vibrations.

An analysis by other researchers concludes that any lever explanation of the increased atlatl dart launch speed is at most incomplete [16]. The transfer of energy from atlatl thrower to the dart is more complex than a simple lever argument would suggest. The reason is that the thrower's movements during the dart launch (Fig. 4a–e) are complicated as he/she adjusts to generate maximum thrust or to aim accurately. I agree with this because I have seen it elsewhere, in spades: in the next chapter we will be looking at slings and there we have much more of the same thing. Nevertheless it is still possible to make simple and meaningful models of thrower motion that permit conclusions to be drawn concerning likely launch speeds, ranges obtained, etc. Just because reality is complex does not mean that (approximate) solutions are invalid, so long as we are careful in choosing the right approximations.

It takes skill and practice to throw an atlatl dart well. (The atlatlist of Fig. 5 maybe needs to practice more.) The throwers must direct the force they apply to the dart in very nearly the same direction as the dart is pointing—if not, the dart will flip either backward (if the force direction is too low) or forward (if too high). They must keep the force and dart velocity in the same direction for the duration of the throw. This is especially difficult to do right for long

[5] See, e.g., the YouTube video *Valley of Fire atlatl event 2014*.

Fig. 5 Warrior ornament of the Moche people, northern Peru, sixth–seventh centuries CE. The dart and atlatl of this gilded copper figure seem to have become detached. Image courtesy of the Metropolitan Museum of Art (NY) Open Access policy

darts, because they buckle considerably during the throw due to the applied force.

Dart flex reduces the speed imparted to the dart, but it does not affect aim, if the atlatlist is skillful (see Fig. 6) and has a perhaps surprising benefit of improving accuracy—by mitigating the torque that arises due to any mismatch between the direction of F and that of v_{atlatl}. See [14] for details.

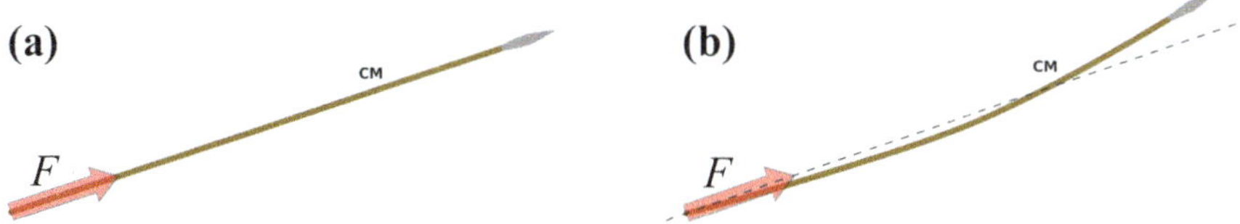

Fig. 6 Atlatl flex effects. A skillful atlatlist directs the force F of his throw along a line to the dart CM. **a** For a stiff dart, this force is directed along the dart shaft. **b** For a dart that buckles under the force, the CM is subjected to the same force and so moves the same way as the stiff dart. But the flex reduces speed by diverting energy from the CM motion

4 Atlatls in Warfare

By "warfare" I mean the large-scale battles that occur between organized antagonists, at least one of whom has provided written records for posterity and, in particular, for fellow humans like you and I who are interested in the development of technology. Of course we can expect that atlatls were used by prehistoric people to hunt fellow humans as well as mammoths: by individuals to settle personal squabbles, by groups skirmishing to decide territorial issues, and so forth. There is not much we can learn about atlatls from these violent encounters, however, that we cannot infer from archaeological records of hunting.

We already know that atlatls can be thrown a sufficient distance and with sufficient accuracy to strike a mobile prey, such as deer, and to strike such target with sufficient momentum to penetrate them to a fatal depth. But how did the atlatl compare against other weapons? How was it regarded by the person at each end of its trajectory? How was it utilized *en masse*?

4.1 In the New World

We have seen that atlatls were developed widely, and most likely everywhere (except Africa) early humans roamed. In some regions the atlatl persisted until recent centuries as a primary weapon, and its use has been recorded in large-scale warfare. In particular, the Mesoamerican societies—Toltecs, Maya, Aztecs—and some South American people, such as the Inca, used atlatls in warfare. From the records of their wars we can learn new things about atlatls.[6]

There has been a debate conducted within the academic literature over "lithic projectile points"—stone arrowheads. Well, not necessarily arrowheads, and that is the point. There are many chipped and flaked pieces of flint and obsidian in the archaeological records telling us that projectile weapons have been made and used, but to what were these points attached? Arrows or darts? The shafts to which they were attached will have been wooden, and so will have decayed away to nothing over the centuries or millennia. In much of the world, they were attached to both; initially mostly to atlatl darts and then mostly to arrow shafts. Atlatl use persisted in the New World (Figs. 1, 2, and 3). A recent paper suggests that in Brazil, they were likely attached to arrow shafts [17] and so it is not there that we look further. Inca records certainly show that atlatls

[6] This is the way that this book will go, in reporting the various projectile weapon history of the ancients. First we will survey widely to estimate the breadth of use of a given weapon, and then home in on one or two regions where it was widely used or where it became specialized. For atlatls it is Central America, for crossbows it is China, for boomerangs...well, you can guess.

were used, but not as the primary weapon (they preferred bows and especially slings). In North America it was a similar story: atlatls were widely used, but bows dominated.

Artistic depictions and archaeological findings suggest that atlatls were a significant weapon of war in Toltec society. These people occupied the Yucatan Peninsula and surrounding areas of Mexico in the seventh to eleventh centuries CE. We are getting warm. Mayan records and iconography depict the atlatl more prominently than the bow—to them it was a ritualized symbol of power, associated with war. A prestige item, often wrought in copper, the atlatl was used from the pre-classic to post-classic times, say from 2000 BCE to 1521 CE (the arrival of the Conquistadors). Even more so for the Aztecs, the atlatl was the principal projectile weapon in warfare, and was associated with elite warriors and magical powers. Bingo—we will see what can be gleaned by investigating the role of atlatls in Aztec warfare.

4.2 In Aztec Warfare

Pre-Columbian Aztecs recorded many aspects of their civilization in codices (bark or animal-skin documents containing pictogram images). Most, but not all, of these codices were destroyed during the conquest, but the few that survive provide primary-source details of Aztec life. The codices, with written text, continued into the colonial era. Apart from these we have Spanish records of the conquest of the Aztec Empire and the manner in which Aztecs conducted war.

"*Montezuma had two houses full of every sort of arms, many of them richly adorned with gold and precious stones. There were shields great and small, and a sort of broadswords ... set with stone knives which cut much better than our swords ... There were very good bows and arrows and double-pointed lances and others with one point, as well as their throwing sticks, and many slings and round stones shaped by hand ... and there were artisans who were skilled in such things and worked with them, and stewards who had charge of the arms.*"—Bernal Díaz del Castillo, 1492–1584 (Hassig [18], p 2.). Castillo was one of the Conquistadors, and participated in the war between Cortés and Montezuma[7] (nowadays called Moteuczomah Xocoyotl).

Between these sources and others we know quite a lot about the use of atlatls ("throwing sticks") by the Aztecs. The weapon itself often had a shaft made of oak, decorated with feathers at the butt end and with a flint, obsidian, copper, or

[7] Fought between two peoples whose civilizations and world-views contrasted astonishingly. Perhaps their similarities, rather than their many differences, can teach us more about humanity as a whole. One thing they had in common was this: both slaughtered millions in the name of their religion. Just saying.

fishbone point at the business end. Some dart points were two-pronged; others were barbed. Spanish sources attest to the effectiveness of atlatls in combat: they could pierce any armor and penetrate deep enough within to cause fatal wounds, from a distance of 50 meters or so. The darts had a penetrating force exceeding that of Aztec arrows at the same range.

Within the armories of the Aztecs, and deployed in battle alongside atlatls, were bows and slings (*tematlatl*), as we have learned from Castillo. The bows were simple self-bows (see Chap. 4) made from wood from the tepozan tree and strung with sinew, with arrowheads similar to the points of atlatl darts. The slings were made from maguey (agave) fiber and shot rounded stones. Castillo mentions the deadly effect of the arrows but reckoned that sling stones were more damaging, due to the number that rained down on their targets. Yet the atlatl was the most effective and most prestigious projectile weapon.

In battle, the Aztec warriors (a high-status example is represented in Fig. 7) would open up against enemy formations with slingstones and arrows, at long range, and then launch atlatl darts when closing in for close-range combat. (In the melee they used wooden "swords" inset along their length with obsidian—sharper than Spanish steel, according to Castillo.) We can guess why the atlatl was used in this way (like the Roman *pilum* or other javelins launched by their infantry against enemy infantry formations just prior to closing with them). At such close range it could be aimed accurately and would penetrate most effectively. Also, launching an atlatl dart required just one arm—a real advantage. (An Aztec warrior used his right arm for throwing, with the left holding a "sword" or war club.) A bow required two arms to fire, and a sling required too much space for closely packed infantry to deploy. The atlatl, after launching its dart, was then quickly stowed or discarded in favor of the warrior's close-combat weapon.

5 Summary

Atlatls were developed in the Stone Ages by many people and were likely their second projectile weapon (after the javelin). Thus they were the first machine developed for projecting missiles. They remained in use for millennia, which speaks to their utility, until being replaced in most (but not all) parts of the world, probably by bows and arrows. As a weapon of war, the atlatl was most significant in Mesoamerica before and during the Conquest.

The atlatl is a lever in the sense stated earlier—it extends the distance over which the dart is subjected to the thrower's force. However dart flexibility and the human element of atlatl throwing, in particular, the customized move-

Fig. 7 Aztec warrior, with atlatl, darts, and shield. (Thermoluminescence date of casting core, 1345–1575). *Courtesy* Cleveland Museum of Art. https://www.clevelandart.org/art/1984.37. Access date April 16, 2025

ments that each thrower makes to optimize the dart motion make the analysis complex, leading throwers up a steep learning curve to acquire skill and leading researchers to make (carefully chosen) approximations.

References

1. See the World Atlatl Association website, https://worldatlatl.org/. Accessed 11 Mar 2025
2. J. Whittaker, Atlatl vs bison on the prehistoric prairie. Rootstalk **10**(1), (2024). Spring. https://rootstalk.grinnell.edu/issues/volume-x-issue-1/

3. See the Encyclopedia Britannica online entry *spear-thrower*, https://www.britannica.com/technology/spear-thrower
4. S.J. Fiedel, *Prehistory of the Americas* (Cambridge University Press, Cambridge, 1987), p. 66
5. B.A. Kipfer, *Dictionary of Artefacts* (Blackwell, Oxford, UK, 2007) p298
6. E. McClellan III, H. Dorn, *Science and Technology in World History* (Johns Hopkins University Press, Baltimore, 2006), p. 11
7. J. Whittaker, D.B. Pettigrew, R. Grohsmeyer, Atlatl dart velocity: accurate measurements and implications for Paleoindians and archaic archaeology. PaleoAmerica **3**, 161–181 (2017)
8. D.L. Lutz, *The Archaic Bannerstone: Its Chronological History and Purpose From 6000 B.C. to 1000 B.C.* (Newburg, David L. Lutz, 2000)
9. H.E.L. Prins, The atlatl as a combat weapon in 17th century Amazonia: Tapuya Indian warriors in Dutch Colonial Brazil. Atlatl **23**, 1–3 (2010)
10. J.R. Ferguson (ed.), *Designing Experimental Research in Archaeology* (University Press of Colorado, Boulder, 2010), p. 199
11. R.B. Lyons, Atlatl to bow. Atlatl **17**, 12 (2004)
12. M. Lombard, Reconsidering the origins of Old World spearthrower-and-dart hunting. Quat. Sci. Rev. **293**, 107677 (2022)
13. N. George, Is Australian flora unsuitable for the bow-and-arrow? Econ. Bot. **78**, 258–273 (2024)
14. M. Denny, Atlatl internal ballistics. Phys. Teach. **57**, 69–72 (2019)
15. R.A. Baugh, Dynamics of spear throwing. Am. J. Phys. **71**, 345–350 (2002)
16. C. Lepers, J. Coppe, V. Rots, The propulsion phase of spear-throwers and its implications for understanding prehistoric weaponry. J. Archaeol. Sci. Rep. **59**, 104768 (2024)
17. M. Okumura, A.G.M. Aranjo, Contributions to the dart versus arrow debate: New data from Holocene projectile points from southeastern and southern Brasil. An. Acad. Bras. Ciênc. **87**, (2015). https://doi.org/10.1590/0001-3765201520140625
18. R. Hassig, *Aztec Warfare: Imperial Expansion and Political Control* (University of Oklahoma Press, Norman OK, 1988)

2

Slings: The Goliath Weapon

Chapter Summary The ubiquity of slings in antiquity, and their variety and that of their projectile, is discussed. Slings were cheap to make but difficult to use; their role in ancient warfare (particularly in Iron-Age Britain) is considered. Different styles of throwing sling bullets are presented and analyzed. A simple formula predicting sling bullet launch speed in terms of slinger power, sling length, and projectile mass is presented. The physical principles underlying sling dynamics (lever action, double pendulum) are analyzed. Sling bullet shape is discussed and the high spin rates that are imparted to these bullets during launch are predicted and compared with data obtained from modern slingers.

"*Whether 'tis nobler in the mind to suffer the slings and arrows of outrageous fortune, or to take arms against a sea of troubles, and by opposing end them?*"—W. Shakespeare (Hamlet, Act III, Scene 1).

1 Slings Were (almost) Everywhere

There is archaeological evidence that Neanderthals knew how to make cord 50,000 years BCE [1], and evidence of a Paleolithic Homo sapiens rope-making tool [2]. Direct evidence of cord from so long ago is rare, of course, because it is a perishable substance, but enough evidence exists for us to be certain that cord has been around forever—more or less. How long did it take to get from cord to a sling projectile weapon? My guess is not long, and the evidence for slings[1]

[1] Here and in the literature *sling* applies to the projectile weapon of Fig. 1. The word *slingshot* refers to the Y-shaped weapon powered by stretched cords. It is also used by NASA to describe the acceleration of a satellite as it whips past a planet, though the physics there is closer to that of a sling.

M. Denny, *Slings and Arrows*,
https://doi.org/10.1007/978-3-032-08563-4_2

goes back to the Stone Ages (see the timeline in the Introduction). The oldest surviving slings are from Peru (4500 BCE); the oldest artistic representation of them is from the Neolithic city of Çatalhöyük, in modern Turkey (9000 BCE) [3].

Unlike atlatls of the last chapter and boomerangs of the next, slings had to be invented from scratch, in the sense that the material used to make them does not exist in the wild; it had to be invented first—pre-assembled from plant material. A natural crooked stick could make an atlatl, and a different crooked stick could make a boomerang (thought the latter required much fine-tuning to get it to fly, as we will see), but very little in nature looks like a practical, usable sling. The sling, I therefore claim, represents a technological step forward, and a mental leap made by somebody (a hunter): "an atlatl can throw a dart but not a rock—what do I have to do to an atlatl so it can throw a rock?"

Not one person made this mental leap—very likely many people did, given that slings were widespread across the prehistoric world, on every continent occupied by humans, except Australia. Slings (Fig. 1) remained an important weapon for hunting and warfare for longer than the atlatl, in most parts of the world, for reasons that we will examine. In at least two times and places, it was indeed a principal weapon of war (for the Inca and for the Celts of pre-Roman Britain).

Whenever and wherever writing was developed around the world, the use of slings was recorded. Thus a Roman historian might tell us not only of the use of slingers in the Roman army, but also of the sling warfare of Rome's enemies. Across the Old World we find that Assyrians, Baleares, Britons, Egyptians, Greeks (Achaens, Athenians, Minoan-Mycenians, Rhodians), Irish, Israelites, Persians, Phoenicians, Romans, Sumerians, and others all made use of the sling

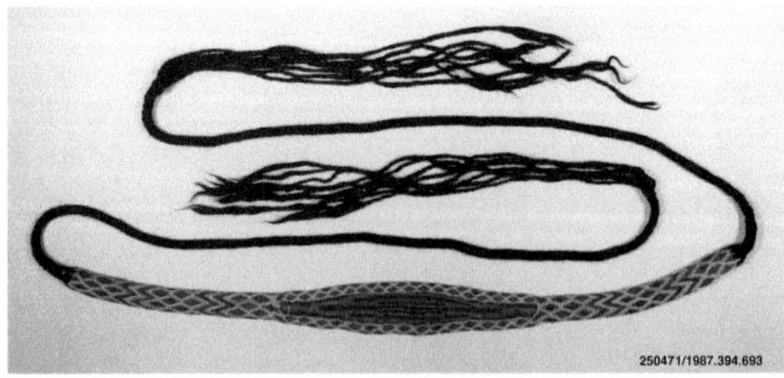

Fig. 1 Inca sling, fifteenth–sixteenth centuries. Each cord length is about 25 inches. Camelid hair (likely guanaco, the most common wild Camelid in Peru, the Inca region.) Some Inca slings were twice this length. Image courtesy of the Metropolitan Museum of Art NY Open Access policy

in warfare [4–6, 23]. It was also widely used across Asia [7], Mesoamerica, South America (Andean civilizations), and Oceania [8, 9]. Wherever ancient people hunted or fought one another (except Australia) there were slings and slingers.

Sling use fell out of favor at different times in different places. It survived as a weapon of war in the New World and the Near East until the eighteenth century [10]. In the Old World it was superseded sooner (in medieval times or earlier) by the bow, for reasons we will see in the next section. It was ubiquitous because easy to manufacture, and cheap. Nevertheless there was a price to pay: the sling was difficult to use properly, and it took many years of practice to become expert.

I need to introduce some terminology at this point, to assist exposition and analysis—this will become a common introductory feature of later chapters as our ancient projectile weapons become more complicated. The projectile fired by a sling is generally called a sling bullet or *bullet* for short. This is true whatever form the missile took, be it a round stone picked up off the ground or a manufactured projectile. The slinger keeps hold of the *retention cord* and lets go of the *release cord* when slinging the bullet. Sometimes these cords are not the same: the retention cord might have a knot to help the slinger keep hold, and the release cord might have tassles, which act as a baffle so that aerodynamic drag slows it down post-release.

2 Why Slings?

So slings can throw rocks harder and further than the strongest man can, but why bother with them at all, except maybe for scaring off predators who are after your sheep? (The sling is sometimes called the "shepherds's sling" for this reason. See the box ***Shepherd vs. Hun***.) Surely a projectile with a point is more effective; so once a group of ancient people have developed pointed projectile weapons, why continue with stone-throwers? There are good reasons.

2.1 Sling Capabilities

The prominent role played by slings in ancient and medieval warfare has long been debated by experts [10, 13]. Yet today the sling is largely forgotten, except by professional historians—the rest of us recall the David and Goliath story and relegate slings to biblical times and places. People today who have not forgotten the sling nevertheless underrate it, even those with an interest in historical weapons. However, in the words of one expert, Chris Harrison:

"*...in experienced hands, the sling was arguably the most effective personal projectile weapon until the 15th century, surpassing the accuracy and deadliness of the bow and even of early firearms.*" [10].

Shepherd vs. Hun

The sixth-century Byzantine historian Procopius wrote *The Persian Wars* in about 550 CE, and tells us the following curious anecdote that is interesting on several levels. From Book 2 Sect. 26:

"*And in the following year, Chosroes, the son of Cabades, for the fourth time invaded the land of the Romans, leading his army towards Mesopotamia ... Chosroes ... uttered a threat in the palace that he would make slaves of all the inhabitants of Edessa and bring them to the land of Persia, and would turn the city into a pasture for sheep. Accordingly when he had approached the city of Edessa with his whole army, he sent some of the Huns who were following him against that portion of the fortifications of the city which is above the hippodrome, with the purpose of doing no further injury than seizing the flocks which the shepherds had stationed there along the wall in great numbers: for they were confident in the strength of the place, since it was exceedingly steep, and supposed that the enemy would never dare to come so very close to the wall. So the barbarians were already laying hold of the sheep, and the shepherds were trying most valiantly to prevent them. And when a great number of Persians had come to the assistance of the Huns, the barbarians succeeded in detaching something of a flock from there, but Roman soldiers and some of the populace made a sally upon the enemy and the battle became a hand-to-hand struggle; meanwhile the flock of its own accord returned again to the shepherds. Now one of the Huns who was fighting before the others was making more trouble for the Romans than all the rest. And some rustic made a good shot and hit him on the right knee with a sling, and he immediately fell headlong from his horse to the ground, which thing heartened the Romans still more.*" Chosroes I (also spelled Khosrau, Khusro, or Khosrow), traditionally known by his epithet of Anushirvan ("the Immortal Soul"), was the Sasanian King of Kings of Iran from 531 to 579. Edessa is modern Urfa in SE Turkey.

Slingers from classical antiquity were credited with great accuracy; thus in the first-century CE the Roman historian Livy wrote: "*Having been trained to shoot through rings of moderate circumference from long distances, they [Greek slingers] would wound not merely the heads of their enemies but any part of the face at which they might have aimed*" [11]. Such claims abound from this period of history (see, e.g., the Bible: "*Among all these soldiers there were seven hundred select troops who were left-handed, each of whom could sling a stone at a hair and not miss.*"—Judges 20:16).

A modern novice slinger has reported hitting a human-scale target (1.15 m high × 0.45 m wide) with a golf ball thrown from a sling seven times out of ten (7/10) at 15 m range, 4/10 at 30 m, and 2/10 at 60 m [12]. Chris Harrison has written much about sling use in history, and has given me any number of historical quotes regarding their accuracy; his opinion is that an experienced slinger from classical antiquity could regularly hit a head-sized target at 50 meters.[2] Lewis Everett of Archaic Arms, an experience slinger who has provided me with much sling data, can reliably achieve 5/10 hits on a 0.5 m diameter circular target at 20 m range. He says the main limitation is lack of uniformity of his stone bullets. He thinks that the ancient Balearic slingers were much more skillful than he is, because they started young.[3]

There is plenty of evidence to suggest that sling bullets used in battle inflicted serious or fatal injuries, even on soldiers who wore armor. It is worth quoting from some classical sources to emphasize this point: "*There it remained fixed while the thongs were whirled round and taut, but when at the moment of discharge one of the thongs was loosened, it left the loop and was shot like a leaden bullet from the sling, and striking with great force inflicted severe injury on those who were hit by it,*" (Polybius, Histories, 27.11.6—*ca.* 130 BCE)—note the reference to lead bullets. Again: "*The inhabitants of the Balearic Islands are said to have been the inventors of slings, and to have managed them with surprising dexterity, owing to the manner of bringing up their children. The children were not allowed to have their food by their mothers till they had first struck it with their sling. Soldiers, notwithstanding their defensive armor, are often more annoyed by the round stones from the sling than by all the arrows of the enemy. Stones kill without mangling the body, and the contusion is mortal without loss of blood*" (Vegetius: *De Re Militari* "Concerning Military Matters," *ca.* 450 CE). Modern forensic estimates of the trauma inflicted by sling bullets phrase their findings in more technical medical language, but the gist is the same [14].

We might perhaps have suspected something about the accuracy of sling bullets, and their potential for serious injury, by recalling that slings had been used as hunting weapons since prehistory. Not only were slings accurate (in expert hands, it should be emphasized) and dangerous, but they also were capable of launching some of their bullets to ranges exceeding a quarter mile, as we will see when we get to a physical analysis of these weapons.

---

[2] Professor C. Harrison, personal communication, April 25, 2025.

[3] L. Everett, personal communication May 6, 2025. We will encounter Archaic Arms and Lewis's sling data again in this chapter, and again in Chap. 8.

Thus we have in the sling a weapon that is comparable in range to the bow, with as projectile inflicting injuries comparable in severity to those inflicted by an arrow. (The sling bullet was harder to defend against that an arrow, however [15], presumably because it was harder to see in flight.) So why did the bow eventually dominate the battlefield? We will have to wait until Chap. 4 to see in detail the capabilities of bows, but we can understand here and now why they bested the sling in battle. First, though, note that because of their similar ranges and accuracy (and also comparable rates of fire) the bow and the sling were used in a similar manner in warfare, with one major exception.

This similarity makes slings and bows analogous to two animals competing in the same ecosystem: one will win out, driving the other extinct. Perhaps, then, a minor difference will decide the issue? One such minor difference between the two is the nature of the injuries that their projectiles caused. Sling bullets impacted an enemy causing blunt force trauma, whereas arrows penetrated. Yet we have seen that sling bullets can be deadly, and we know (if we watch movies or read Chap. 4) that arrows are also deadly. So, this difference did not drive bows to dominance in warfare and slings to (near) extinction. A major difference was that bows gave rise to a whole new class of warrior in the ancient world: the horse-archer—we will meet these tough nuts in Chap. 4. Here I will say only that they certainly helped spread the use of the bow, but did not see off the sling from battlefields of old.

On those battlefields, rows of archers and/or rows of slingers fired at enemy infantry formations from behind a protective wall of their own infantry. (Bowmen and slingers did not carry shields, and so would be very vulnerable if they formed the front row.) More commonly the slingers would be used for skirmishing, which is to say that they would form a loose array in front of enemy formations—close enough to direct missiles upon their enemy but far enough away, sufficiently fleet of foot, and sufficiently dispersed that they were not vulnerable to enemy projectiles or cavalry attacks. Slingers had to be dispersed compared to close-order ranks of infantry, because they needed room to operate, and so this skirmishing role was where they excelled. Archers *could* form close-order ranks and did so in many an ancient battle, as we will see, and this is the reason (IMHO) why bows eventually became the standard projectile weapon of medieval armies. Thus a line of soldiers that is, say 300 yards across might consist of 200–300 archers, but only about 50–100 slingers; thus, all else being equal, the archers could rain down a greater number of projectiles upon an enemy formation. The reason for bows' success is packing density—geometry—not any particular superiority of the weapon. Packing density and horse archers—but the latter will have to stay put, no doubt impatiently, on their pasturelands until Chap. 4.

Fig. 2 Cypriot lead sling bullets, from classical antiquity. Length 2.9–6.5 cm. Image courtesy of the Metropolitan Museum of Art NY Open Access policy

2.2 Costs and Benefits

The sling was easy and inexpensive to make; compared with the other projectile weapons in this book it was the cheapest, perhaps alongside a low-end atlatl. Boomerangs were more expensive in labor, if not materials. Bows were more-to-much-more expensive in both labor and materials. Crossbows more so again, in materials. Siege engines—well, as you may imagine, these were the largest and certainly the most expensive projectile weapons of old.

Sling bullets were free, at least in their most basic form of rounded pebbles. Slingers learned, however, that they could sling a bullet further if it were made of lead, and more accurately if it had a bicone or spheroid shape (see Fig. 2). Lead bullets were developed quite early,[4] say 500 BCE.

The extra range, compared with slinging stones, was due to the increased density of lead, which meant that a projectile of given weight would have less volume and so less area subjected to aerodynamic drag. The typical spheroidal bullet would have a long axis about twice that of the other axes, and when released it would spin about this long axis. Such a bullet would be spin-stabilized and so would be more accurate. (We will look into ballistic details shortly, as part of our investigation of sling physics.) Not only that, but the spinning bullet would fly through the air with its long axis pointing in the direction of movement. This means that the drag force, which is proportional to the projected area presented to the atmosphere through which it passes, is as small as it can be. (Thus a spheroidal lead bullet would be subjected to lower

[4] Some lead sling bullets have been found on the site of the battle of Marathon, 490 BCE, though they cannot be definitively linked to the battle. More certain is a 2023 find of a lead bullet with Julius Caesar's name on it [16].

Fig. 3 A Balearic slinger (ca. 200 BC) with three slings for short-, medium- and long-range targets. Here one sling is worn as a headband, and another as a belt. Public Domain image by Johnny Shumate

aerodynamic drag force than would a spherical lead bullet of the same weight.) It is these small lead bullets which could be shot a quarter mile, as we will see.

The motion of a bullet while it is within the pouch of the sling, prior to release—the internal ballistics trajectory—is in general quite complicated, as the slinger has to accelerate it and ensure that it is lined up with the target at the instant of release. Because of this complexity it is not easy to learn how to use a sling. The difficulty of mastering the sling meant that it takes some years of practice to become expert, and so historically it tended to become a speciality. Indeed, in classical antiquity slingers from the Balearic Islands[5] fought as mercenaries for the Carthaginians and later for the Romans, having a reputation within both these empires for their skills with the sling (recall the quote from Vegetius). See Fig. 3. Note that our Balearic slinger possesses three

[5] Which include the islands of Menorca, Mallorca, and Ibiza, today tourist hot-spots.

slings (long, short, and in between—for long, short, and in-between ranges, with different types of bullet and perhaps different style of launching it, for each).

Now we turn to sling science—a tricky but rewarding subject, because it turns out that we can simplify the complex movement of the slinger and sling and bullet, and obtain useful estimates for sling performance that match up with observations of sling action.

3 Sling Physics

There are many hobbyists out there who are fanatical (that *may* be too strong a word—let me downgrade it to "keen to use and learn") about slings.[6] Many are interested in the history of slings; they make and use slings which are good, bad, or indifferent replicas of ancient Balearic slings (or Greek or Inca or ...) and they test them. If we can assume that a modern hobbyist slinger is about as adept at using a sling (well, maybe not quite) as a slinger from classical antiquity, and if we can assume that they are using similar bullets, then these hobbyists can tell us something about the capabilities of ancient slings.

There's a lot of "ifs" in the last sentence, so we must be wary of simply adopting modern results that have been obtained for sling performance. Thus, Roman and Greek historians have made great claims for the ranges attainable by slingers of their era. Are these claims credible? Well, modern slingers can throw small, lead bullets a quarter mile [10] but does that mean the ancients could do so? More than likely, but we should check. Recall that we have already checked the claims of old about the lethality of sling bullets against modern medical reports. The same needs to be done for ancient claims about sling range. The results we obtain by physical analysis will show what is possible and, when added to the data obtained by hobbyists hurling sling bullets, we can be pretty confident of our conclusions. We can check not only range capability in this way (by matching up theory with practice) but also bullet spin rate and slinger accuracy.

3.1 Styles of Throwing Sling Bullets

I will say it again and not for the last time: the movement of a slinger, and so of the sling bullet, during the launch phase is generally very complicated; because

[6] There are many keen hobbyists who like to talk about, read about, design, make, and use all of our ancient projectile weapons. I will have more to say about them in the final chapter.

of this it is easier to picture the motion by watching YouTube videos that by reading about it. I will therefore, in this section, frequently refer you to such videos (most of which were created by hobbyists for other hobbyists, but which are nevertheless very useful for our purposes of "experimental archaeology").

The most basic method for you to launch a bullet from a sling is this: hold your sling arm up and behind you, as if you were about to throw a stone. The sling hangs vertically. You then throw the sling, letting go of the release cord at just the right moment so that the bullet heads off in the right direction. Perhaps only beginners use this method—it provides some of the benefit of a sling (your effective arm length is extended, so the bullet launch speed is increased somewhat) but includes no rotation which can be used to build up speed. Which brings us to the so-called "Greek" style, as seen in depictions on ancient Greek coins.[7] The pouch-holding hand is stationary, pointed toward the target, while the hand holding the sling is near the slinger's head, resembling an archer's stance. Then the sling is swung three-quarters of a rotation (270°) before release. This style has longer range than the basic method just described because the bullet is accelerated for a longer time before being released and so the launch speed is greater.

The simplest style of throwing with a sling is what I will call the "two-handed pirouette" style: it resembles the action of an Olympic hammer thrower. The sling is long and the bullets used for this style are heavy; from ancient sources we think the bullets weighed about one pound. The sling in this case can be up to eight feet long (each cord). The slinger holds one cord in each hand and rotates their body with increasing speed; they are always facing the bullet at this stage, not looking at the target, and the bullet accelerates as the slinger turns faster and faster. Typically three turns were made before releasing the bullet. Because of the increased number of turns, and because the sling is longer, the two-handed pirouette style can throw a bullet much further than the Greek style (but the rate of fire is lower). I call this the simplest style, but this is a physicist's perspective—we will soon see that it is the simplest to analyze, though not the simplest to master.

There is also the "one-handed pirouette" manner of throwing a sling bullet, sketched in Fig. 4. The slinger could aim more easily if they rotated the sling above their head without rotating their body as well—that way they could keep a steady eye on the target. This way is indeed the manner of most one-arm styles, but here the sling is too long and the bullet too heavy—the slinger must hold the sling out to one side and rotate with it, as for the two-handed pirouette. The slinger carries out at least one full rotation while accelerating

[7] See Chris Harrison's *Slinging.org* website for a description of this, and other slinging styles.

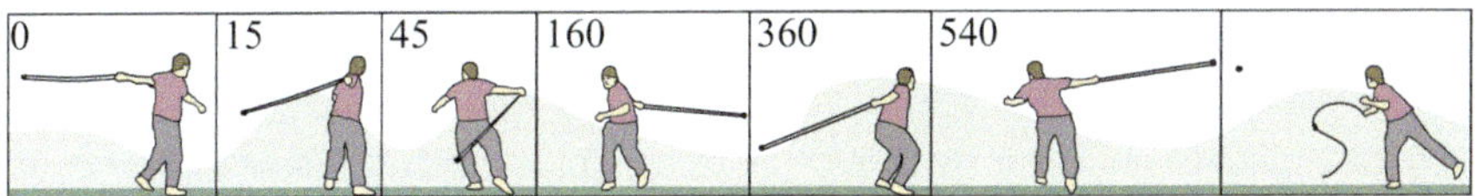

Fig. 4 One-handed pirouette slinging style. In this case the slinger rotates his body and the sling about $1\frac{1}{2}$ revolutions before releasing the bullet. Numbers refer to the approximate sling rotation, in degrees

Fig. 5 Side-arm style. From the left: initial rotation (only the right wrist, and sling, move), acceleration (x 3), launch

the bullet. The one-handed pirouette style throws a bullet further than the Greek style, but not as far as the two-handed pirouette. It makes it easier for the slinger to move forward as well as rotate, as you can see from YouTube video demonstrations[8] Yet another manner of throwing a sling bullet has been called the *whip* style; it was popular in Southeastern Asia and in Oceania, and is described by Richardson [17].

I will use the term "side-arm" to describe a whole class of one-handed styles, in which the slinger rotates a short- or medium-length sling lazily above his head (or to one side) while picking out a target or waiting for the right moment. Then they whip their sling arm around with increasing speed and release the bullet. The acceleration part of this motion may cover only half a rotation, or maybe more. Side-arm sling movement is what most people first think of when told about slings. A very good example of this style can be seen in the slow-motion YouTube video *Balearic slinger in competition (slo-mo)*; in this example, the bullet is accelerated for $1\frac{1}{2}$-2 rotations prior to release. The stages of this style are sketched in Fig. 5.

[8] See, e.g., *Slinging Pirouette Style Tutorial (for power)*, and *Ancient Long Sling Hit 35 m.*

3.2 Sling Range

How was it possible for a slinger of ancient times to throw a bullet over 400 yards when he or she would have struggled to throw it without the sling a quarter of that distance? We know that it is possible from modern sling trials[9] but it would be good to know that the science can back up these results (and predict maximum possible range—spoiler: the modern record is 437 m or 478 yards [10]).

In TA-2 you will find a derivation of the following equation for the launch speed of a sling bullet:

$$v \approx \left(8\pi r \frac{n\bar{P}_n}{m}\right)^{1/3}. \tag{1}$$

Here r is the radius of the bullet rotation (arm length plus sling length), n is the number of rotations through which the bullet is being accelerated, $\bar{P}_n$ is the average power that the slinger transfers to the bullet during the acceleration phase, and m is bullet mass. Once we have determined the launch speed, then we know that the maximum range is given by $R = \eta \frac{v^2}{g}$ (assuming a 45° launch angle) where g is the (constant) acceleration due to gravity at the earth's surface and η is a number less than one which tells us how much effect aerodynamic drag has on projectile range. From TA-0 we obtain the range with and without drag for a given initial launch speed v and so find that, for example, $\eta \approx 0.44(0.66)$ for a 50-g stone (lead) bullet with $v = 60$ ms^{-1} (200 fps), or $\eta \approx 0.28(0.48)$ for $v = 90$ ms^{-1}.

We must unpack the algebra of Eq. (1) before putting in some numbers to see what kind of launch speeds and ranges are possible. So r is a constant for the pirouette-style sling throws but not for the side-arm style because, for the latter, arm length increases throughout the acceleration phase. Also, the sling and arm are not generally aligned during the launch for side-arm throws (there is an angle between them that can change). So for side-arm throws we replace r by $\bar{r}$, the average value. The number n of full 360° rotations varies with style, and with slinger. It is important to remember that n refers to the phase when the bullet is being accelerated around a circle (when power $\bar{P}_n$ is being transferred from the slinger's arm) and does not include the initial twirling that slingers seem to go in for at the beginning of their throw—that twirling is at constant rotation speed and so the amount of bullet angular acceleration is little to none.

[9] See, e.g., the YouTube videos *Slinging for Distance—Day 1, Smith Creek, NV* and *Slinging for Distance—Day 2, Smith Creek, NV*. Note that the slinger refers to sling bullets as *glandes*; this is the term Romans used—it means "acorns."

I need to discuss the average power $\bar{P}_n$ in a bit more detail. It refers to the mechanical power that is transferred to the bullet, and not to the power expended during the throw by the slinger. The latter includes the power of muscles moving arms and torso (and moving the slinger—who often runs forward during the throw) and is several times larger than the mechanical power. (Experiments show that about 25% of the metabolic energy expended by muscles is converted to mechanical energy [18, 19].) We can estimate the mechanical power without needing to know the hard-to-measure biophysical power by comparison with other mechanical activities. Say you lift a 50-lb sack of potatoes in one second from the ground to a bench that is $3\frac{1}{2}$ feet high. You have imparted gravitational energy to the sack (by raising it) at an average rate of $\bar{P} = 200$ W (watts[10]). This rate of energy transfer is quite easily achieved by a fit person, and could be repeated every minute or so for quite some time before they got too tired. So we can reasonably suppose that a slinger might impart energy to a bullet at a similar rate. Let us say that for a side-arm slinger the average rotation radius (arm plus sling, recall) is $\bar{r} = 5$ feet, the slinger's action is such that the bullet is accelerated over $n = 1\frac{1}{2}$ rotations, and the lead bullet weight is 2 oz. For these parameters we substitute into Eq. (1) and find that our slinger can send the bullet on its way with a launch speed of about 58 ms^{-1}. For such a speed, the maximum range is about 230 m (250 yards), allowing for aerodynamic drag. A two-hand pirouette throw of a 1 lb stone bullet (for a 2.5 m cord length—8 ft 4 in—and $n = 3$ rotations) with the same slinger power transfer will reach a launch speed of 43.6 ms^{-1} and a range of about 170 m. These are quite impressive numbers.[11]

To obtain the current record for a sling bullet throw, and to get close to some of the historical claims, our slingers need to up their game. From the modest power of 200 W let us say that they put their backs into it (and torsos and legs) and increase the power transfer rate to 500 W (say, two kilowatts of metabolic power) acting for one second. This rate is equivalent to carrying that 50-lb sack of potatoes up ten steps in one second—still doable, but you wouldn't want to have to repeat it more than once every five minutes or so. At this increased power, all else being equal, the two-ounce bullet will travel 390 m if the slinger can manage $n = 2$ rotations, and the one-pound bullet will fly 220 m (240 yards).

[10] Power is rate of change of energy with time. Yes, I know I am mixing units here, but feet and watts are commonly used together in most parts of the English-speaking world.

[11] I can imagine that both a Roman soldier and his helmet would have been impressed by a 1-lb rock striking them from over 180 yards away.

The 2-oz bullet range is getting up toward the current world record. If longer ranges are to be achieved then either more power is needed, or bullets with lower aerodynamic drag.

Equation (1) holds for a wide variety of sling actions, and this is because the manner in which the slinger's power accelerates the bullet doesn't matter—it may be a constant value throughout the throw, or it may increase throughout the throw, but so long as the average power $\bar{P}_n$ is the same then the launch speed is given by Eq. (1). (Approximately, as always—a number of simplifying assumptions went into the calculation. See TA-2.)

3.3 Levers and Double Pendulums

Equation (1) tells us that bullet launch speed increases as slinger power increases, as effective radius increases and as the number of rotations increases. Launch speed falls as bullet mass increases. None of this is a surprise. We have shown that quarter-mile-plus ranges are possible—but why? What physical principle permits ranges so much greater than the distance a rock of the same weight could be thrown by hand? First and simplest: the lever principle. The slinger's effective arm length l is increased by the length of the sling to r, say. So, for a given rotation speed that he can generate, the bullet speed is proportionally faster than the rock speed (the bullet speed at any point of the acceleration phase is greater than the rock speed by a factor of about r/l).

Second is the so-called *double-pendulum effect.* Non-physicists know this phenomenon better as the *slingshot effect* by which satellites in space whip past a planet, gaining energy from it, flying away much faster (and in a different direction). The term is unfortunate because it brings to mind a slingshot—the Y-shaped weapon beloved of schoolkids plinking at tin cans or the neighbor's cat. The physics has nothing to do with that type of slingshot.

A simple example of the double-pendulum effect is shown in Fig. 6, for the special case with the angles a, b are the same and changing at the same rate ω. Ponder this figure for a minute and you will see why the double-pendulum bob moves faster. The analysis of double-pendulum motion is much more complicated than that of a simple pendulum (the single pendulum on the left of Fig. 6), but the figure conveys all that you need to know. Think of the double pendulum as a golfer with arms of length A and a golf club of length B, pivoted at the golfer's wrists.

An arm (length l) plus a side-arm sling is a double pendulum of total length r, and so the speed of the bullet can be made to be greater than that of an extended arm of length r. It is important to understand that there are two effects here, both of which increase sling bullet speed. First the sling extends arm length

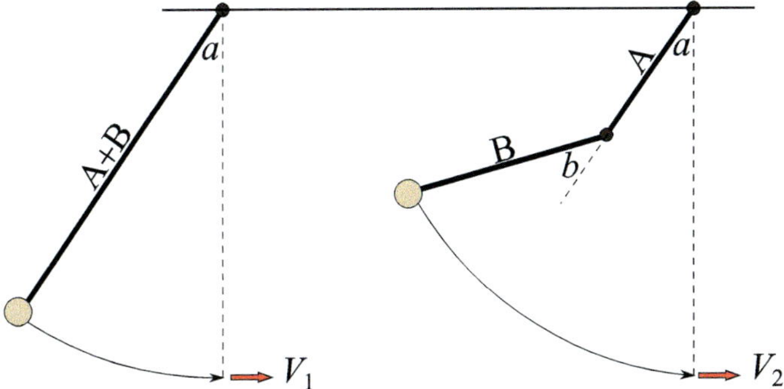

Fig. 6 Double-pendulum effect. Both angles a and b are decreasing at an angular rate ω. For the single pendulum the bob speed at the bottom of the swing is $V_1 = (A + B)\omega$, whereas for the double pendulum $V_2 = (A + 2B)\omega$

and so, by the lever action, increases speed. Second, the double-pendulum effect increases speed further. We will return to the double-pendulum effect in Chap. 7, because the subject of that chapter—siege engines—was often equipped with slings.

Note that for the two-handed pirouette style, there is no double-pendulum effect. However in that case the lever effect is greater because the arm extension is greater—such slings can have eight-foot cords, whereas side-arm sling cords are usually three feet long, or less.

3.4 Bullet Spin

There are two aspects to sling bullet spin that we will address: spin rate and spin alignment. How fast does a bullet spin upon release, and in what direction? Does this matter? Yes, big time. Why?

Let us begin by addressing the spin rate. The reason why a bullet acquires a spin when it is released by a sling is clear enough, and can be seen at a glance from Fig. 7.

While the bullet is being accelerated by the slinger, during the internal ballistics phase, it is subjected to (centripetal) forces from both cords of the sling that keep it in place. At the instant of release the force due to the release cord ceases but the retention cord force remains, as shown in Fig. 7b. A very short time (call it δt) later the bullet is entirely free of the sling and is flying toward its target. During that short interval there is a one-sided force F acting on the bullet which applies a torque and so generates spin. In TA-2 I obtain

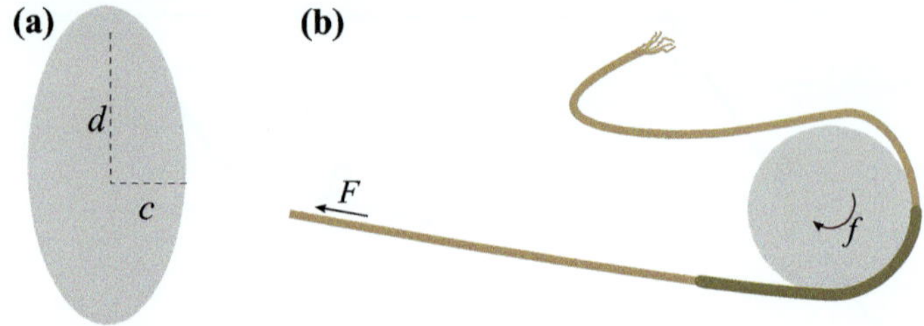

Fig. 7 Bullet spin. **a** A spheroidal bullet, with one elongated axis. **b** Seen end on, immediately after release the bullet is subjected to a force F due to the pouch and the retention cord. The release cord is free. After a brief interval δt the bullet separates from the sling

an expression for that spin in terms of other bullet parameters:

$$f \approx \frac{v}{\sqrt{2\pi r c}}. \tag{2}$$

In words, the bullet spin frequency is about equal to the bullet launch speed divided by the square root of the product of bullet circumference and bullet minor axis radius. Here I have assumed that the bullet of Fig. 7 is a spheroid with a circular cross section of radius c and with one elongated axis of length d. Typically for sling bullets, the long axis is about twice the short axis, $d = 2c$, and this is what I have assumed in the derivation of Eq. (2).

Let us assume that the bullet has a launch speed of $v = 45\ \text{ms}^{-1}$ (call it 150 fps). A two-handed pirouette sling with rotation radius $r = 8$ feet and a 1-lb stone bullet of radius $c = 1$ inch will be thrown a maximum distance of about 180 yards, and will be given a spin of about 70 Hz (4,200 rpm). A side-arm sling with rotation radius $r = 4$ feet and a 2-oz lead bullet of radius $c = \frac{1}{3}$ inch will have a range of up to 250 yards, and will set out on its trajectory with a spin of 230 Hz (13,700 rpm).

These spin rates seem extraordinarily large, and reflect the large forces. There is some spin data out there against which we can compare these numbers, but not a lot at the time of writing. Spin rates have been measured by sling enthusiasts to be $f = 45$–70 Hz for tennis ball "bullets" and $f = 60$–250 Hz for small lead and stone bullets.[12] Our estimates are entirely compatible with these numbers, which suggests that the calculations of TA-2 are on the right

[12] For observed bullet spin rates, see, e.g., the YouTube videos *Slinging physics—Release timing and spin influences* (and the associated webpage https://slinging.org/forum/YaBB.pl?num=1703086191/6), and *Slinging spin rate.*

track and that Eq. (2) is a valid reflection of the dependence of bullet spin frequency upon its speed and radius, and upon sling rotation radius.

What about spin alignment? In what direction is the spin axis pointing? We would very much like it to point in exactly the same direction as the bullet velocity, for two reasons: increased accuracy and increased range.

A bullet spin that is along its velocity direction provides *spin stabilization.* Careers have been devoted to this subject because of its military significance for rifle bullet and artillery shell stability in flight, and trajectory accuracy. The *gyroscopic effect* ensures that, if a bullet is spinning fast enough, then it will not tumble in flight. Tumbling (turning end over end, so that the sides as well as the front of the bullet are exposed to the on-rushing air) is disastrous for both accuracy and range: accuracy because the tumbling causes erratic flight, and range because the average aerodynamic drag increases substantially when the bullet is tumbling.

So how much spin is needed to prevent tumbling and thus to stabilize the bullet flight? There is a widely used formula called *Miller's twist rule*[13] that provides a criterion for the optimum twist rate (number of inches within a rifle barrel for which the rifling completes one turn, or rotation). This rule applies for oblate-spheroidal bullets resembling our sling bullets, and so applies to our case. Miller's rule can be turned around to provide a formula for the *gyroscopic stability factor* S_{Miller}, a dimensionless constant. In TA-2 I show that, for our case, this constant is given by

$$S_{Miller} \approx \frac{0.06\,m}{rc^2}, \tag{3}$$

where m is bullet mass in kilograms, r is sling rotation radius in meters, and c is bullet radius in meters. The bullet trajectory is unstable if $S_{Miller} < 1.0$, is stable if $S_{Miller} > 1.5$, and for S_{Miller} between 1.0 and 1.5 is—meh—(author holds out open hand, palm down, and waggles it back and forth) doubtful. Substituting parameter values into Eq. (3) for our side-arm sling and two-handed pirouette sling, we find that $S_{Miller} \approx 44$ and $S_{Miller} \approx 17$, respectively. Thus the spin that is naturally imparted to a sling bullet is sufficient to stabilize that bullet in flight.

[13] See the Wikipedia site *Miller twist rule* for the formula, or any number of American gun club websites for calculators applying it.

4 Sling Warfare in Pre-Roman Britain

Slings were the dominant projectile weapon in Iron-Age Britain (800 BCE to the Roman conquest of southern Britain in 43 CE). Here seems to be one of those examples where the sling and bow were, if not completely, then to a significant extent mutually exclusive—there is very little evidence of bows in Britain at that time [20, 21]. Unfortunately the Celtic people on the island[14] at the time were not literate and so we do not know their history; much of what we do know comes from archaeology, with a few written records from the conquering Romans at the very end of the period.

It seems that in Iron-Age British society, which was fractured and warlike, the two-man chariot was a significant military weapon (unusually—in most parts of the world at this time cavalry had taken over from chariots). Charioteers were the elite warriors, and the driver could skillfully maneuver their vehicles even through woodlands. They would skirmish like horse-archers elsewhere: the charioteer would rapidly approach enemy formations and the elite warrior would then throw javelins at the enemy before the chariot retreated. I can do no better than let Julius Caesar describe what he saw in battle against British tribes: "*In chariot fighting the Britons begin by driving all over the field hurling javelins, and generally the terror inspired by the horses and the noise of the wheels are sufficient to throw their opponents' ranks into disorder. Then, after making their way between the squadrons of their own cavalry, they jump down from the chariot and engage on foot. In the meantime their charioteers retire a short distance from the battle and place the chariots in such a position that their masters, if hard pressed by numbers, have an easy means of retreat to their own lines. Thus they combine the mobility of cavalry with the staying power of infantry; and by daily training and practice they attain such proficiency that even on a steep incline they are able to control the horses at full gallop, and to check and turn them in a moment. They can run along the chariot pole, stand on the yoke, and get back into the chariot as quick as lightning*" (Gallic War, IV.33).

Most British warriors carried a spear and shield, according to the artwork that survives from this period. There is remarkably little evidence of slings or of the use of slings and as a consequence the role of slings in Iron-Age Britain was understudied until recently [21]. So how do we know that slings were the dominant projectile weapon in warfare? It seems that they dominated siegecraft. Those chariots would quickly dispatch or disperse any slingers that would in other armies act as skirmishers in front of the enemy, and so slings were

[14] I will follow common practice and use the term *Britain* to refer to *Great Britain*, the largest of the *British Isles*: a geographical term referring to an archipelago off the northwest coast of Europe, including the second largest island, *Ireland*.

Fig. 8 "An aerial view of Maiden Castle in Dorset from the west." Photo taken by Major George Allen (1891–1940) October 16, 1937

mostly used attacking, and certainly defending, the many hill forts that dotted the British landscape at the time—some 3,300 have been indexed, and most of these were from the Iron-Age period [22]. The largest of these was Maiden Castle in Dorset, southern England, expanded from 16 acres to cover 47 acres around 450 BCE. Some 22,000 sling bullets were found during excavations of this site, the ramparts of which survive (see Fig. 8). Many others were found at other hill forts and elsewhere [23]. British sling bullets were biconical or spheroidal, as elsewhere, but unlike those used by the Romans they were made of stone or clay.

The interesting aspect of many of these hill forts is that they appear to have been designed specifically for sling warfare.[15] Maiden Castle was a contour fort encircling a hilltop. It was *multivallate* (more than one defensive rampart or enclosure) with the inner ramparts higher, so that defenders there could throw their sling bullets over an outer rampart at besieging enemies. There were different styles of hillfort but many were complex with this feature—staggered or interleaved ramparts, with zig-zag entrances not adjacent to those of the next rampart inside. There were sling platforms for defenders to rain missiles down on besiegers, and well-planned lines of fire.

[15] This view was widely accepted, almost universally so, until a generation ago when some historians walked back a little from it and suggested that some castles had other purposes, perhaps with permanent settlements inside the larger ones such as Traprain Law in southern Scotland. Perhaps some were used for ritual purposes or as seasonal meeting places [22].

With their interesting offbeat military structure (chariot skirmishers, forts built for sling warfare) the Celts of Iron-Age Britain provide a tantalizing glimpse into the significance of slings in the conflicts of prehistory.

5 Summary

Slings are very old weapons, easily constructed from readily available natural materials. They are difficult to use, however, and it takes years of practice to become very skillful at throwing sling bullets. Slings match bows for range, accuracy, and rate of fire, and for millennia both were used around the world by many people, though there is evidence to suggest that, to some extent, they were mutually exclusive. In the Old World sling use declined in the Middle Ages, as the bow was more suited to the type of military activity then predominating.

The motion of a sling bullet during the internal ballistics phase, when it is being accelerated by the slinger but prior to its release, is in general three-dimensional and subjected to many different and varying forces. Also it is subjected to torques so the bullet when launched has spin as well as velocity. To fly true, the spin must be aligned with the velocity direction; this occurs naturally for the pirouette style of throwing but not for other styles, which require precise actions by the slinger to ensure.[16]

References

1. B.L. Hardy, M.H. Moncel, C. Kerfant et al., Direct evidence of Neanderthal fibre technology and its cognitive and behavioral implications. Sci. Rep. **10**, 4889 (2020). https://doi.org/10.1038/s41598-020-61839-w
2. N.J. Conard, V. Rots, Rope making in the Aurignacian of Central Europe more than 35,000 years ago. Sci. Adv. **10**, eadh5217 (2024). https://doi.org/10.1126/sciadv.adh5217
3. C. Fitzgerald, How sling weaponry revolutionized warfare in the ancient world. War History Online (2021). Accessed 26 Apr 2025
4. E.C. Echols, The ancient slinger. Class. Wkly. **43**, 227–230 (1950)

[16] The last paper I had published prior to writing this book was about sling physics. As mentioned in the Introduction physics journals are understandably hesitant about publishing papers with significant historical content, yet my sling paper was accepted, and even highlighted by the American Journal of Physics. Why? Due to the very complications just summarized; to obtain tractable equations it was necessary to make significant approximations, and this interested the editors. The primary result was Eq. (1) which permits sling launch speed to be estimated from known input parameters. Its validity—despite the approximations—has been confirmed by comparison with previously published sling data [24].

5. G. Haklay et al., Up in arms: slingstone assemblages from the late prehistoric sites of 'En Zippoi and 'En Esur. 'Atiqot **111**, 1–22 (2023)
6. L. Keppie, *Slings and Sling Bullets in the Roman Civil Wars of the Late Republic, 90–31 BC* (Archaeopress, Oxford, Uk, 2023)
7. R. Dohrenwend, The sling-Forgotten firepower of antiquity. JAMA **11**, 28–49 (2002)
8. A.R. Recinos, O.A. Firpi, R. Rodas, Evidence for slingstones and related projectile stone use by the ancient Maya of the Usumacinta river valley region. Anc. Mesoam. **33**, 309–329 (2022)
9. R. York, G. York, *Slings and Slingstones-The Forgotten Weapons of Oceania and the Americas* (Kent State University Press, Kent OH, 2011)
10. C. Harrison, The sling in medieval Europe. Bull. Primit. Technol. **31**, 74–79 (2006). Harrison discusses the ranges of ancient as well as modern slings
11. Livy, The History of Rome. Book **38**, 29.6
12. D. Jackson, The Iron Age shepherd sling. EXARC J. Issue 2019/4, https://exarc.net/ark:/88735/10459. Publication date 2019-11-25
13. M. Korfmann, The sling as a weapon. Sci. Am. **229**, 34–42 (1973)
14. I. Borovsky et al., The traumatic potential of a projectile shot from a sling. Forensic Sci. Int. **272**, 10–15 (2017). https://doi.org/10.1016/j.forsciint.2016.10.006. (Epub 2016 Oct 26 PMID: 28088089)
15. W.K. Pritchett, *The Greek State at War: Part V* (University of California Press, Berkeley, 1991), p. 56
16. J.M. Ordax et al., En torno al Bellum Hispaniense y las Glandes Inscriptae de Hispania. Un nuevo proyectil con inscripción cesariana procedente de Montilla (Córdoba). Zephyrvs **91**, 183–195 (2023). https://doi.org/10.14201/zephyrus202391183195
17. T. Richardson, The Ballistics of the Sling. R. Armouries Yearb. **3**(1), 44–49 (1998). https://doi.org/10.1080/30650682.1998.12426625
18. J.C. Dean, A.D. Kuo, Energetic costs of producing muscle work and force in a cyclical human bouncing task. J. Appl. Physiol. **110**, 873–880 (2011)
19. R. Margaria et al., Energy cost of running. J Appl. Physiol. **18**, 367 (1963)
20. D. Harding, *Iron Age hillforts in Britain and Beyond* (Oxford University Press, Oxford, UK, 2012), p. 194
21. D. Swan, Attitudes towards and use of the sling in late Iron Age Britain. Reinvention: An Int. J. Undergrad. Res. 7, (2014). https://warwick.ac.uk/fac/cross_fac/iatl/research/reinvention/archive/volume7issue2/swan/. Accessed 26 Apr 2025
22. A.H.A. Hogg, *British Hill-Forts: an Index* (B.A.R, Oxford, UK, 1979)
23. S.J. Greep, Lead Sling-Shot from Windridge Farm, St Albans and the Use of the Sling by the Roman Army in Britain. Britannia **18**, 183–200 (1987)
24. M. Denny, Internal ballistics of the sling. Am. J. Phys. **93**, 367–375 (2025). https://doi.org/10.1119/5.0226263

3

Non-Returning Boomerangs: Propellers from the Paleolithic

Chapter Summary The ancient origins of boomerangs are outlined. The complicated aerodynamics of boomerang flight are simplified, for ease of presentation and also to provide insight. Returning boomerang trajectories are due to precession motion. The historically more common hunting (non-returning) boomerangs are discussed and their idealized equations of motion derived. The fine-tuning of these boomerangs to straighten the trajectory is presented in some detail. Boomerang flight stability requires a bent or curved shape.

"The war boomerang is an effective and dangerous weapon, having a range of 150 yards, and having been known to pass completely through an adversary when the body was first struck by the point of the weapon." (Blackman, 1903) [1].

1 A Roundabout Introduction to Boomerangs

Perhaps you are a little surprised to learn that there *are* such things as non-returning boomerangs. (After all: "*Of the popular fallacies associated with the boomerang perhaps the most widespread and deep-rooted is the belief that all boomerangs are of the returning type. As a matter of fact, returning boomerangs constitute only a very small percentage of Australian boomerangs, a percentage difficult to estimate accurately but which under normal aboriginal conditions may have been exceedingly small*" Davidson, 1935 [2].) Indeed they were the vast majority of historical boomerangs because, unlike their return-to-sender brethren, they had practical use. The returning boomerang (RB) was regarded by the aboriginal people of Australia as a toy—for use in play and sporting

M. Denny, *Slings and Arrows*,
https://doi.org/10.1007/978-3-032-08563-4_3

Fig. 1 "Aborigines of Lake Tyers," Charles Bayliss, 1887. Digital image courtesy of Getty's Open Content Program

competitions—whereas the non-returning boomerang (NRB) was a serious multi-purpose tool as well as a cultural artifact (Fig. 1); principally a weapon for hunting medium to large game and for war. It is an undisputable fact that this finely tuned piece of military hardware represented, for a period of at least 20000 years, the leading edge of human aerodynamics technology. (This is the reason why boomerangs appear in this third chapter; they are scientifically more advanced than spear-throwers or slings, but less so than bows. The technology of boomerangs is on a par with the subject of Chap. 1—they are shaped sticks—but is so much more refined.)

Boomerangs are free-flying propellers. They rotate as they fly and this is what gives them aerodynamic lift, and why they can have so many different shapes—see Fig. 2. The wing shape of a boomerang—be it a sporty returning boomerang or its hard-working straight-shooting cousin—is not as important as its airfoil shape (the cross section of the wings).[1] A note on technical terms: for any boomerang (say a classic V-shaped RB) the straight parts are called *arms* (or *wings*—we will use both words) and are joined at the *elbow*. The *dingle arm* is the one held for throwing (a.k.a. the trailing wing) and the other, leading wing is the *lifting arm*. It is perhaps surprising that the shape does not even have to be symmetrical. The reason is rotation: if the rotation rate is high enough then, aerodynamically, the boomerang performs very similar to a circular disk of the same size (that is, the area traced out by a boomerang spinning about its

[1] "*A fact too little recognized is that the shape of the cross section can be more essential to a boomerang's flight than the precise shape of its platform*" (Hess [3], p. 14).

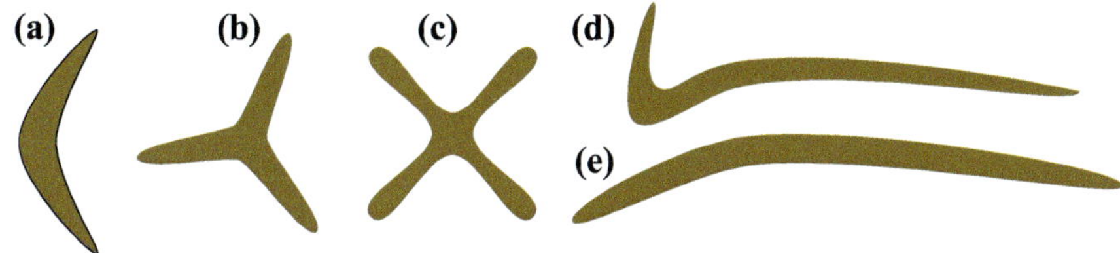

Fig. 2 Boomerang forms. **a** Traditional returning boomerang. **b** and **c** Modern RBs. **d** Traditional hooked or beaked NRB. **e** Traditional non-returning boomerang, typically larger and straighter than RBs

center[2]), if it spins at a fast enough rate. How fast? My calculations suggest a few cycles per second—easily generated by a natural throwing action, perhaps enhanced in some cases with a deft flick of the wrist at release.

The arms are shaped to provide just the right amount of aerodynamic lift and to minimize aerodynamic drag. The cross sections need to be flat underneath and curved on top. The boomerang is twisted along its length, like a propeller. If done more or less right, the RB flies in a curved arc. If done oh-so-very-right, tuned with exquisite precision, then the NRB flies straight.

The approximation described here—describing a boomerang as aerodynamically like a disk of the same radius because of its rotation—is known to aerodynamicists as the *actuator disk model.* It can be a good approximation, but it is not exact. If you are a perfectionist and require the best approximation that physics has to offer—exact solutions do not exist in the real world—then you must perform the full CFD (computational fluid dynamics) calculations, which require detailed knowledge of the boomerang wing shape. These calculations apply the Navier-Stokes equations of fluid dynamics which are notoriously difficult to solve, and especially so for boomerangs because of the complicated boomerang motion, which involves rotation about a center as well as movement through the air. The leading edge of one wing provides lift, but after a fraction of a second the boomerang rotates so that this leading edge becomes a trailing edge, and so its aerodynamic properties change—lift and drag forces vary along the wing, and vary with the rotation angle. (For those readers who crave Reynolds-averaged Navier-Stokes solutions to a blade-element theory of boomerang dynamics, see [4, 5]. For those who do not, read on.)

For reasons that will become clear, boomerangs are right-handed or left-handed. That is to say, boomerangs that are to be thrown by a right-handed thrower are not the same as those intended for a southpaw. The two

[2] Technical detail: the center of rotation is the same as the center of mass if the boomerang is symmetrical, but they are not necessarily the same for asymmetric forms.

boomerangs are mirror images (this is true for both RBs and NRBs, and for all shapes and sizes). Returning boomerangs are thrown with the plane of rotation nearly vertical (see, for example, Perry's instructive video [6]) whereas NRBs are thrown with the plane horizontal [7]. It is interesting to note that the radius of a returning boomerang flight is insensitive to the initial speed, initial rotation rate, and layover angle[3]—it is a fixed property of the boomerang. For our approximate actuator disk model we find that the radius of the flight trajectory increases with boomerang weight and, perhaps surprisingly, decreases with boomerang lift coefficient and size (see TA-3). This result is well known and is borne out by observation; it gives me confidence to apply the actuator disk model to the much-less-studied NRB flight trajectory.

There is a rather careful description by a nineteenth-century observer of boomerang throwing by a native Australian, which conveys something of the skill required, so it is worthwhile to provide it in full: "*When a skillful thrower takes hold of a boomerang with the intention of throwing it, he examines it carefully (even if it be his own weapon, and if it be a strange weapon still more carefully), and, holding it in his hand, almost as a reaper would hold a sickle, he moves about slowly, examining all objects in the distance, heedfully noticing the direction of the wind as indicated by the moving of the leaves of trees and the waving of the grass, and not until he has got into the right position does he shake the weapon loosely, so as to feel that the muscles of his wrist are under command. More than once as he lightly grasps the weapon he makes the effort to throw it. At the last moment, when he feels that he can strike the wind at the right angle, all his force is thrown into the effort: the missile leaves his hand in a direction nearly perpendicular to the surface; but the right impulse has been given, and it quickly turns its flat surface towards the earth, gyrates on its axis, makes a wide sweep, and returns with a fluttering motion to his feet. This he repeats time after time, and with ease and certainty. When well thrown, the furthest point of the curve described is usually distant one hundred or one hundred and fifty yards from the thrower*" (Smyth, 1878 [8]).[4] (Much of the trajectory description here is predicted by the actuator disk model for an RB.)

Boomerang physics had been empirically understood (mastered, or controlled, would perhaps be better words) by pre-scientific cultures for thousands of years but in the post-Renaissance Western World they were unknown—forgotten relics of a distant past. Boomerang dynamics could not be understood by science until scientists learned of the existence of these whirling toys.

[3] Layover angle is the angle that the boomerang makes to the vertical. It is small (a few degree) for RBs and is 90° for NRBs.

[4] Curiously, historical sources claim that native Australians always seemed to throw their boomerangs with the free (lifting) arm facing forward, though this should make no difference to the trajectory [3, 9].

And it was the toys that fascinated them—the lightweight frivolous RBs and not the heavyweight, serious, straight-shooting weapons.

In the early nineteenth century the British Empire stretched from Australia all the way back to the United Kingdom, which in those days included Dublin, Ireland. It seems that the boomerang also found its way to the old country and became an object of popular fascination, much like Rubik's cubes or frisbees in the twentieth century: "*Of all the advantages we have derived from our Australian settlements, none seems to have given more universal satisfaction than the introduction of some crooked pieces of wood shaped like a horse's shoe, or the crescent moon; and called boomerang, waumerang or kilee. Ever since their structure had been fully understood, carpenters appear to have ceased from all other work; the windows of toy shops exhibit little else; walking sticks and umbrellas have gone out of fashion; and even in this rainy season no man carries any thing but a boomerang; nor does this species of madness appear to be abating*" (Dublin University, [10] 1838, p168). The same article that provided this light-hearted observation of a popular craze also provided the first scientific explanation of how RBs work. The anonymous author may have known of the then recent developments of George Cayley who spelled out the role played by the four forces acting upon a flying object: lift, drag, thrust, and gravity. He or she[5] will have been familiar with the earlier works of Newton, Bernoulli, and Euler, but not with later work crucial to a detailed understanding. Nevertheless a correct qualitative explanation was given: the boomerang does not fall, due to a lift force generated by rotation; the trajectory is circular due to *precession*, the same physical phenomenon that is responsible for the wobble of spinning tops.

With these boomerang basics under our belt, we can dispense with the RB toys and devote the rest of this chapter to the NRB tools and weapons. We will first delve into their prehistory and history before learning how to fell a kangaroo or deer or bad guy with one.

2 Paleolithic Origins

We need to back up a little, to set our non-returning boomerang into historical and geographical context, before getting to grips with NRB physics and engineering. Here we look into the long and widespread history of boomerang evolution. It is perhaps to be expected that the development and refinement of boomerangs reached its peak in the world's oldest surviving culture, in Australia, though boomerangs were invented independently in many parts of the world.

[5] Unlikely, because women were not admitted to Dublin University until 1904.

In particular there is recent evidence of an evolution of throwing sticks in Europe dating back to the Paleolithic. Indeed according to one detailed study there is, surprisingly, more archaeological evidence for throwing sticks/boomerangs in Stone-Age Europe than in Australia, while there are far fewer ethnological records [11]. The universal lack of physical remains is to be expected, whereas other projectile weapons such as arrows and spears leave stone tips for archaeologists ten millennia down the line to discover, while slings leave sling bullets, the throwing stick does not leave anything. The paucity of any kind of recent record in Europe is also to be expected given that other projectile weapons came to the fore across all of the Old World. In Australia a lack of competition in the field of projectile weapons led to the persistence of throwing sticks and their eventual evolution into boomerangs, so that post-Neolithic archaeology reveals more evidence of boomerangs than is found elsewhere in the world.

So, going forward in this chapter we will certainly mention instances of boomerangs in other parts of the world, but will concentrate on Australia. Just bear in mind that the first part of the journey happened on other continents as well—very likely on all continents where early humans lived and hunted.

The first lesson we learn from the extensive archaeological record is that there is a continuum of development from ordinary *stick* to returning boomerang. That is, sticks along with stones will undoubtedly have been the earliest missiles used by humans; throwing a stick at a nearby rabbit might have resulted in a meal. Sticks of a certain size will have been better than others for this purpose, and sticks with a circular cross section will not have flown as well as those will a flatter cross section. So now we have a *throwing stick.* (There is evidence for throwing sticks going back 300,000 years [12].) Over time (centuries, millennia, epochs) hunters noticed that a carefully shaped surface helped the throwing stick go further, and carving the leading edges helped even more. Now we have superior throwing sticks—here I will call them *flightsticks*—they fly further than other throwing sticks because their shape conveys some aerodynamic lift. Over more time hunters found that fluting (see Fig. 3) or polishing the flattened surfaces of their flightstick aided flight. They now saw that each flightstick and each thrower were different, and that warping the flattened flightstick in a certain way caused it to fly off in a curved trajectory that brought the stick back to the thrower. ("Hmmph—not much use for hunting, but the kids think it's fun.") And it can bring down some birds in a big flock. Further fine-tuning, by carving or twisting or scalloping the surface differently at different points, caused the flightstick to fly straighter and further,

Fig. 3 Hooked boomerang with fluting along its length. Thanks to Arthur Beau Palmer for permission to reproduce this image from his collection

with ranges getting over 100 yards [13]. Now we have our true non-returning boomerang (a *kylie* or a modern American *rabbitstick*).[6]

The finished hunting boomerang flew very straight which made it easier than a rock or an arrow to aim at a target, for two reasons. First, aerodynamic lift meant that the thrower did not have to aim upward to counter the effects of gravity; just aim straight at the target—the stick will counter gravity on its own. Second, these NRBs were bigger than returning boomerangs, $2\frac{1}{2} - 3$ feet in length, and so the thrower's aim could be off by a foot or more and the target still gets hit. (Recall that NRBs are thrown so that they rotate in a

[6] "*The boomerangs are manufactured from greenwood cut to the desired shape and angle. The points are hardened by drying in hot sand or ashes, after which the weapon is bent to the required twist while held firmly on the ground by the ball of the foot and wrenched with the hand. But even after this treatment the boomerang is not finished until repeated trials of its flight have been made, and it is chipped, scraped and twisted until its working qualities are considered perfect*" (Jennings and Hardy, 1899) [14].

horizontal plane.) Being big and heavy and hard, when they hit they kill or incapacitate.

The Old Stone Age ended around 12,000 years ago. Well before this transition to the New Stone Age (Neolithic—see the timeline shown in the Introduction[7]), when humans were beginning to learn about agriculture and build permanent dwellings, they had learned to make certain tools and weapons—and one of these early weapons was the boomerang. The earliest remains of a boomerang date from 23,000 years ago: in 1986 a well-preserved and near-complete boomerang carved from a mammoth tusk was unearthed in the Obłazowa cave in the Tatra mountains, southern Poland [15]. The shape is that of a sickle with both ends pointed, the length is $2\frac{1}{2}$ feet point to point. Crucially for identifying this artifact as a boomerang is the cross section: carved flat on one side and curved on the other, like an airfoil. Also it is twisted, though this might have occurred naturally over millennia.[8]

Another problem archaeologists face in definitively identifying partial remains as boomerangs is that a shard with an airfoil-like edge, for example, might just be that shape by chance—the stick it came from might not be shaped to provide a lift force all along its length. Despite these limitations enough remains have been found, in very varied circumstances, to show that boomerang-like throwing sticks have evolved—presumably independently—on all the continents of Earth that were occupied by humans. That is, many steps along the continuum from stick to fully fledged hunting boomerang were made by many different cultures.

The oldest boomerang in Australia was found in 1973 in the Wyrie Swamp of the south, and is about 10000 years old. (The oldest record of Australian boomerangs, however, is twice that age: the Bradshaw/Gwion rock art paintings of the Kimberley region in northern Australia date from 20000 years ago.[9] See Fig. 4) About the same age, as determined by radiocarbon dating, is an archaeological find in the American southwest. The National Museum of the American Indian has a number of throwing stick/boomerang artifacts on exhibit, dating much more recently, and used by Hopi-Pueblo people. In Egypt, the tomb of Tutankhamun contained a collection of 20 boomerangs, some made of ivory

[7] Some anthropologists insert a Mesolithic Stone Age between Old and New, but its dates vary with region and it is dispensed with for most parts of the world.

[8] A recent report suggests that the Obłazowa Cave boomerang may be older than first thought—perhaps as much as 40,000 years [16].

[9] Dr. John Hayward has researched the rock art of Australia and attests to the age of Australian boomerangs. He cannot say whether or not the first people who occupied the continent (perhaps 65,000 years BP) brought the boomerang with them or developed it independently after they arrived. (D. Hayward 2025, personal communication).

Fig. 4 Baiame Cave, Milbrodale, Hunter Valley, NSW, Australia. This rock art is thought to be at least 3,000 years old. Thanks to Sardana for this image.

and some of bone.[10] These are 3,300 years old; older Egyptian depictions of hunting scenes with boomerangs have been found (a stone plate from Abydos is 5000 years old, a painting showing Libyan warriors with boomerangs from the reign of Queen Hatshepsut is 3,500 years old). It seems that boomerangs were valued tools in ancient Egypt, and that they could fly some distance and perhaps even return, indicating that they were closer to true boomerangs than to throwing sticks, on the wide spectrum of these weapons.[11] Throwing sticks with aerodynamic features that suggest they may have flown, dating from the first millennium BC, have been found in several places in Europe. In Asia, the *valari*, an ancient metal throwing stick/boomerang, is iconic for Tamil people of southern India; it is said to be thrown 100 yards [15, 19–22].

How was the first boomerang invented? If not gradually by the mundane evolution of throwing sticks as just described, then maybe as follows. A fanciful notion that can only work in Australia, this purely speculative idea appeals: "*Mr. Hubert de Castella has suggested that the Aborigines derived the invention of the Wonguim [boomerang] from observation of the shape and the peculiar turn of the leaf of the white gum-tree. As the leaves of this tree fall to the ground, they gyrate very much in the same manner as the Wonguim does; and if one of the leaves is thrown straight forwards, it makes a curve and comes back. Such an origin for a weapon so remarkable is not to be put aside as unreasonable. It is very probable*

[10] Note the prevalence of deserts. Boomerangs—true NRBs—work best over fairly flat, open land.

[11] "*I made a facsimile of the Egyptian boomerang in the British Museum, and practised with it for some time upon Wormwood Scrubs, and found that in time I could increase the range from fifty to one hundred paces, which is much further than I could throw an ordinary stick of the same size with accuracy*" (Pitt Rivers [18] 1872).

that if children played with such leaves, some old man would make of wood, to please them, a large model of the leaf, and its peculiar motions would soon give rise to curiosity and lead to fresh experiments" (Smyth [8] 1878). You decide.

3 Keep on the Straight and Narrow

Straight flight is important for a projectile weapon, be the target a kangaroo or a king, as we have seen. Yet we have seen that torques acting on a boomerang cause its path to deviate. The name of the game for boomerang throwers is to carve and twist and polish so as to eliminate all torques (except maybe for the torque that slows down the rotation, because that one does not affect flight direction) so that their NRB will fly very straight.

The straight-flight trajectory of an NRB is unstable—it always wants to veer away to the left or right. Consider a soccer ball rolling along the crossbar of a soccer goalpost; it will fall off to one side or another pretty quickly. Only if the ball is sent in *exactly* the right direction will it stay on the crossbar for any distance. This analogy conveys pretty well the instability of our NRB flight.

We begin with an idealization—unattainable in the real world, but one which tuning strives to reach—by assuming that tuning has been made perfectly; somehow the NRB has been modified so that there are no torques acting that cause the trajectory to veer left or right or up or down. What is the resulting trajectory? How far will the NRB fly? How fast must it rotate? These are the questions that I answer in this section.

For a flying machine such as a jet fighter there are four forces that act on it: thrust pushes it forward, aerodynamic drag pushes it backward, gravity pulls it down, and lift pulls/pushes it up. In our case, the NRB is given an initial forward speed but there is no thrust term. The other three forces are known and from these we can calculate the idealized NRB trajectory (see TA-3 for details). An example is shown in Fig. 5. You are a hunter—perhaps a Narangga aboriginal Australian hunting kangaroo 5000 years ago or a Nevada insurance salesman hunting rabbits or deer last weekend. You release your NRB horizontally at a height of 39 inches over flat terrain. Your NRB weighs 18 ounces and is $31\frac{1}{2}$ inches from tip to tip. It sets off at 34 mph toward your intended dinner, with an initial rotation rate of 120 rpm (or 2 Hz—two cycles per second). On that day the dimensionless aerodynamic lift and drag coefficients are $c_L = 0.071$, $c_D = 0.0004$. The resulting flight trajectory is calculated in TA-3 and plotted in Fig. 5a. You missed your target (but only just): your NRB hits the ground 292

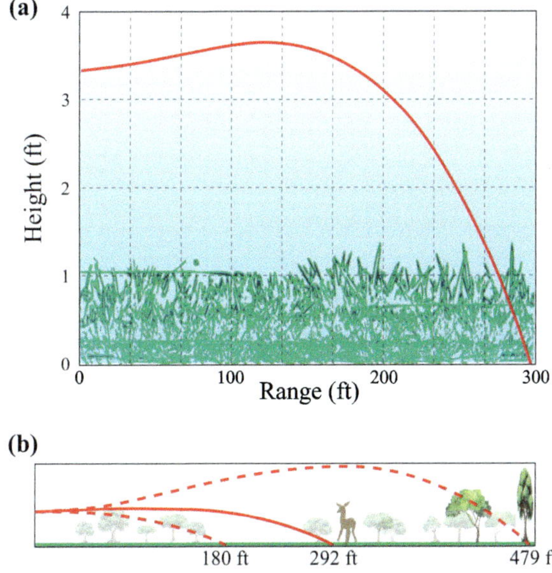

Fig. 5 NRB flight profile, for aerodynamic parameters given in the text, assuming an idealized and perfectly tuned boomerang. **a** Note the nearly level flight for the first 230 ft. **b** Changing the NRB launch speed by $\frac{1}{4}$ mph up or down has a big effect on its trajectory (dashed lines). The flight is also very sensitive to lift and drag coefficients

feet away, 6 seconds after release.[12] Note from Fig. 5a that the boomerang rises a little before descending. This happens because lift force initially dominates over gravity, but then aerodynamic drag slows the rotation rate a little, reducing the lift force. In fact over the trajectory the NRB will lose only about 2% of its initial speed and about 3% of its spin. (It will land with 88% of the energy it started with.)

An interesting result emerges from this analysis: the trajectory is very sensitive to how you throw your NRB. For example, if the initial speed is changed slightly—increased or decreased by just $\frac{1}{4}$ mph—then the trajectory changes as shown in Fig. 5b. If slower by $\frac{1}{4}$ mph then the distance traveled drops to 180 feet, and if faster by $\frac{1}{4}$ mph then it increases to 479 feet with a big increase in height. Both of these will miss the target by a lot. The same result obtains if we alter the lift coefficient by a little—the trajectory changes a lot. (Lift coefficient might change a little day to day with temperature.)

[12] Let us say that your target is 300 feet away, and that your boomerang bounces off the ground and then hits it—so you get your dinner. The historical records show that hunting boomerangs quite often struck their target after bouncing in this manner [3]. Note that boomerangs do not make much noise in flight, and so the target does not hear it until it gets very close.

It has been worthwhile to go through this exercise of calculating an NRB trajectory for the idealized case (of perfect torque elimination via tuning) for two reasons. First, we have learned the kind of ranges that an NRB can attain for realistic input parameters (weight, lift and drag coefficients, initial speed, etc.) and note with satisfaction that these results are entirely plausible, reproducing the kind of ranges observed historically and achieved today by modern boomerang enthusiasts. Second, we have learned that it takes skill to throw an NRB—even an idealized one—so that it will hit its intended target, because the trajectory is very sensitive to the way that the NRB is thrown.

Without the idealization of zero torque it would be much harder to hit a target at distance, because the trajectory would zig-zag all over the place. This, of course, is why throwers must tune their boomerangs—especially NRBs, which are used for hunting. Let us now look into boomerang tuning.

4 Fine-Tuning

We saw in the last section what straight flight looks like, and so have seen the potential capabilities of a NRB. Now we look into what tuning it takes to get there. In practice tuning of a boomerang means altering its surface so that its flying characteristics are adjusted.

Typically an NRB that flies well will be tuned by the thrower so that it flies straighter. Maybe Paleolithic Pete has a boomerang that flies fairly straight for 20 yards and then veers off—it can hardly be classified as an NRB, more like a cross between RB and NRB. Pete wants to increase the straight-flying distance because he can't get to within 20 yards of a kangaroo mob he has his eyes on to feed his family. So he soaks and heats and twists and dries his boomerang to adjust the airfoil shape. He tests the remodeled NRB and finds, after several iterations, that he can now throw it straight for 35 yards. Next he whittles a bit off the leading edge of his boomerang right *here* and another short section just *there*. Then he throws it a few times to see the effects. He whittles a bit more *there*, and tests again. This whittling process is slow and laborious because only a little can be shaved off at any one time—its hard to undo whittling, if a mistake is made, for a wooden boomerang. After much testing Pete finds his range is now 55 yards. Now he polishes parts of the upper and lower surfaces—60 yards. Now he scallops (flutes) the surfaces along their length, *just so*. After more testing, and adjusting slightly his throwing technique for this particular NRB, he finally achieves a straight-flight range of 70 yards. This is the best that he can do with this particular piece of wood.

Here we have an important point: historically most boomerangs were made of wood, chosen from a section of tree branch or trunk that possessed a natural bend. It was carved so that the grain of the wood ran along the length of the boomerang, for strength. Each was unique and so required customized tuning. Throwers such as Pete adjust the tuning parameters of each NRB so that the lift coefficient (those whittlings along the leading edges) vary along the length of each wing. Same for drag coefficient (polishing and fluting). We have seen that the performance of an idealized NRB is sensitive to initial velocity, and Pete will have found the same for each of his boomerangs—so he learned to throw each differently so that the launch speed, spin, and layover were just right to maximize straight-flight range.

Pete and his Paleolithic pals did not know about drag coefficients as modern science knows about them, but they knew that fluting and polishing enabled a boomerang to fly further. Polishing reduces drag by creating a smoother surface, as we might expect. Fluting is less obviously beneficial. We now know that it creates a thin boundary layer of air, a kind of skin over the boomerang surface that stays with it as the boomerang flies, reducing resistance. (Think of a puck sliding over a surface of ice, and then of a puck sliding over an air hockey table with air blown upward through tiny holes in the surface.) In addition to reducing aerodynamic drag, Pete's twisting of the boomerang shape and his whittling of edges changed the angle of attack and the local lift coefficient of his NRB.

Aerodynamicists can reproduce Pete's tuning theoretically, or rather can provide a very simple example of how his tuning can remove the torques which cause the flight path to veer off to one side or to rise or fall. Consider Fig. 6a. Here we see a boomerang that is both flying through the air and spinning, as shown. Clearly the right end of the wing is flying faster through the air than the near end, because of the way that linear speed and rotational speed combine. Now it is well known that the aerodynamic lift force increases with the square of speed (speed of air across the wing surface). So if the lift coefficient is the same at all points along the leading edge of each wing then the lift force generated at each point is as sketched in Fig. 6b. The right wing or arm of Pete's boomerang generates more lift force than the left. But this is no good for a NRB because it causes the boomerang to flip (counterclockwise, for the boomerang of Fig. 6b) at low rotation speeds, or rise (at high rotation speed) due to precession. This is the (counterintuitive) way that precession works: see the box ***Precession*** below for an explanation. So Pete tunes his NRB by differentially whittling—altering the lift coefficient differently at different points along the NRB leading edges so that the coefficients are symmetric (Fig. 6c). Now there is no torque and the NRB will fly straight.

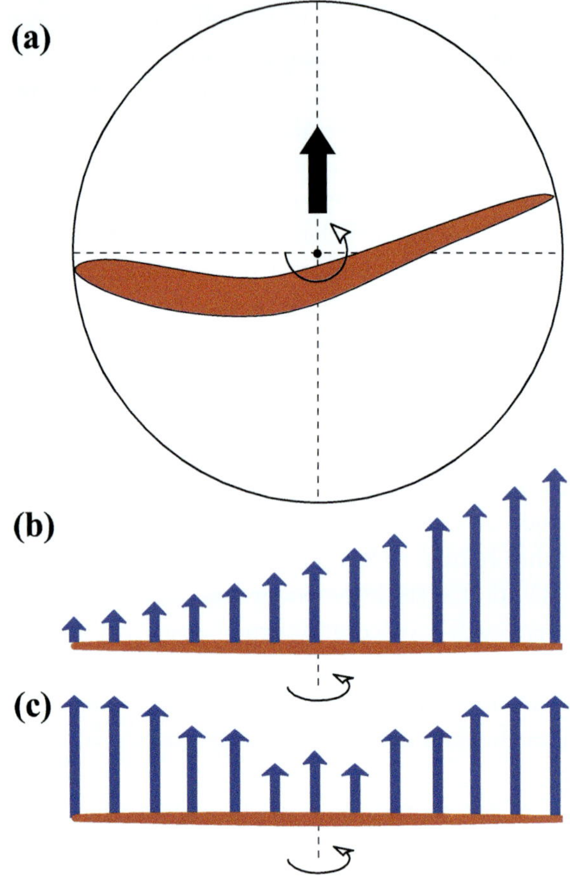

Fig. 6 Tuning aerodynamic lift. **a** Viewed from above, this right-handed boomerang moves in the direction of the arrow, spinning as shown. **b** Viewed from behind (so that the boomerang is flying into the page) the speed through the air is greatest at the right edge of the boomerang and least at the left edge, so the upward lift forces of different parts (blue arrows) are likewise varied. Result: lopsided spin and *torque*—the boomerang wants to flip counterclockwise. **c** When tuned, the lift forces even out. This boomerang does not want to flip sideways—it will fly straight

I can quantify this idea of eliminating unwanted torques to tune an NRB, by showing mathematically how lift coefficient must change so that a torque vanishes—this calculation is outlined in TA-3. This is not the way Pete would have done it, and is indeed a simplification of the problems he had to solve (which had multiple torques), but it is enlightening for math-minded readers [23].

5 Stability: Momentum and Angular Momentum

While it is true that non-returning boomerangs are generally less crooked in shape than their come-to-poppa relatives, they are never straight. Why not? There is nothing in the physics that we have seen so far that forbids straight wings. Some propellers have just two arms, after all, and we are using a propeller model of boomerangs, so what is missing here?

Stability. We saw earlier that boomerang rotation had to be chosen correctly to ensure horizontal flight. Here we see that rotation also provides stability for a wide class of boomerangs. Those that are straight or nearly straight are unstable unless the rotation rate is very high—which is why most NRBs are bent, if only a little. Let's now discuss spin stabilization, after the following public service announcement.

Nerd Alert The last couple of sections have been pretty heavy on the physics, and the next couple will be more so—the math is incarcerated in TA-3 but some of the physics has broken out here. Sorry, but this intrusion is a reflection of the innate complexity of boomerang motion. **End of Nerd Alert**

Utter the words "spin stabilization" in the ear of a physicist and he will instantly mumble the knee-jerk response "gyroscope." There is good reason for this; gyroscopes are the epitome of spin stabilization and embody it explicitly. We have already encountered gyroscopes earlier in this chapter when I introduced (but did not explain) the phenomenon of precession. Spin stabilization is counter-intuitive for people who, by accident of birth, happen not to be geeks. We intuitively understand the addition of forces but not the addition of torques; that is, we understand the vector addition of momentum but not the vector addition of angular momentum. Let's look into this matter a little further—because it is interesting physics—and then see why it gives rise to spin stabilization, after which we will apply this notion to our NRBs to see why they cannot be perfectly straight. (Then and only then will I subject you to an explanation of precession.)

We understand *momentum* as the heft of an object. A cannonball which has twice the mass of another moving at the same speed has twice the momentum. A cannonball with the same mass as another but moving at twice the speed also has twice the momentum. Similarly we understand *force* as the rate of change of momentum: this was one of Isaac Newton's great gifts to the world. His Second Law of Motion states that force is the rate of change of momentum. It takes a certain amount of force to slow down and stop that cannonball (or to get it moving in the first place). It takes twice the force if the cannonball is moving twice as fast or is twice as heavy. Momentum and force are *vectors*—they have direction. Add two equal forces that act in opposite directions and they cancel. Add two equal forces that act perpendicular to each other, and the result is a force at 45° to both.

All this makes sense. But the angular equivalents of momentum and force are not so well understood outside of physics class. Yet angular momentum and angular force (better known as torque) are also vectors and behave similarly.[13] But not identically. Torque is the rate of change of angular momentum, just as force is the rate of change of momentum, and this leads us directly to precession and spin stabilization.

Angular momentum has a direction, but it is not the same as that of momentum. Consider a moving mass. Its momentum is in the direction it is moving. But now suppose that it is spinning on a fixed point, like a gyroscope. It has plenty of angular momentum but zero momentum. What direction is the angular momentum pointing, if the mass is fixed to a point? Imagine a spinning top on a table in front of you. The base is fixed but the mass is spinning. We say that the direction is given by the *right-hand rule*: if you curl the fingers of your right hand in the direction of spin, then your extended thumb defines the direction of the angular momentum. So a top that spins counterclockwise has an angular momentum that points upward.

No need to worry about the details. All I ask is that you grasp the idea that momentum and angular momentum both have a direction associated with them, and that these directions are not necessarily the same. A rotating mass has angular momentum which changes direction when a torque is applied, just as a moving mass changes direction when a force is applied (unless the torque/force happens to be in the same direction as the angular momentum/momentum). A top that is leaning over, and is not spinning, will *fall down* due to the force of gravity, whereas a top that is leaning over and spinning will *fall sideways* due to the way that torque changes the direction of angular momentum (we saw this earlier when discussing precession). The top has been stabilized by the spin, in the sense that it does not fall down. With a very large angular momentum (high spin rate and large mass) the sideways movement is reduced—and with an infinite angular momentum the top doesn't move at all. This is spin stabilization.

[13] Important technical qualification: angular momentum and torque are *axial vectors*, a.k.a. *pseudovectors*, which are almost the same as vectors, except when we consider reflections and other symmetry properties. The interested reader could do a lot worse than consult the Wikipedia article *Pseudovector* for enlightenment on this subject.

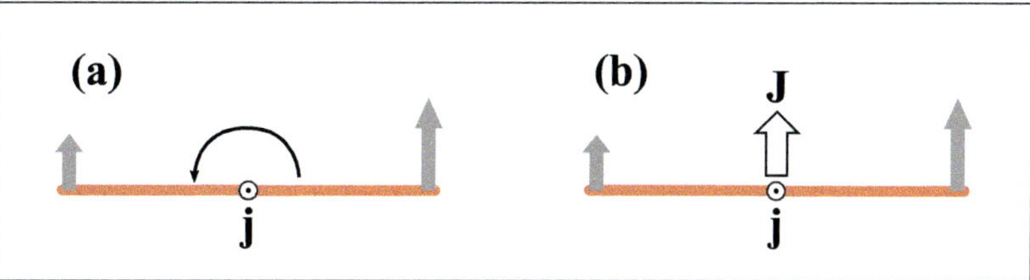

Precession

This is math-lite-precession-101; for the (more concise) mathematical explanation, see TA-3. (a) You throw horizontally a frisbee (thick line—you are looking at it from the back) without spinning it. The frisbee is subjected to different lift forces on the left and right wings—say due to an updraft of wind—shown here by the upward-pointing arrows. The result is a flip (curved arrow)—the frisbee banks left, as you might expect. We say that it has gained a small angular momentum $\boldsymbol{j}$ in this case directed toward you, out of the page (the direction comes from the right-hand rule). (b) You throw the same frisbee in the same updraft, but this time with spin. We say it has an initial angular momentum $\boldsymbol{J}$, directed upward. So now $\boldsymbol{j}$ adds to $\boldsymbol{J}$ so that the frisbee total angular momentum is in the direction $\boldsymbol{J}+\boldsymbol{j}$. If $\boldsymbol{J}$ is much larger than $\boldsymbol{j}$ then the direction of $\boldsymbol{J}$ changes only slowly with time. Again applying the right-hand rule we see that the plane of the frisbee tilts backward and the frisbee climbs. This is precession: a slow change in the direction of angular momentum due to an applied force that is perpendicular to the direction you expect (the non-spinning frisbee banks left, the spinning frisbee rises).

6 Boomerang Stability

With the phenomenon of spin stabilization under our belts, we can now see why an NRB is unlikely to be stable if it is straight—why all real NRBs are bent. (Another reason, I need hardly point out, is that sticks are naturally bent.) Maybe only bent a small angle, or with one arm much shorter than the other, but bent nevertheless. It is easy to understand what can happen if an NRB is as straight as that of Fig. 7 by throwing a wooden ruler—as often as not it immediately sets to spinning about the long axis (the **x**-axis). This instability occurs because it is a lot easier to get the ruler/NRB to spin about the long axis than about either of the other two. To see why, let us suppose that, for the NRB of Fig. 7, the dimensions are such that $a, b << l$ (a and b are much

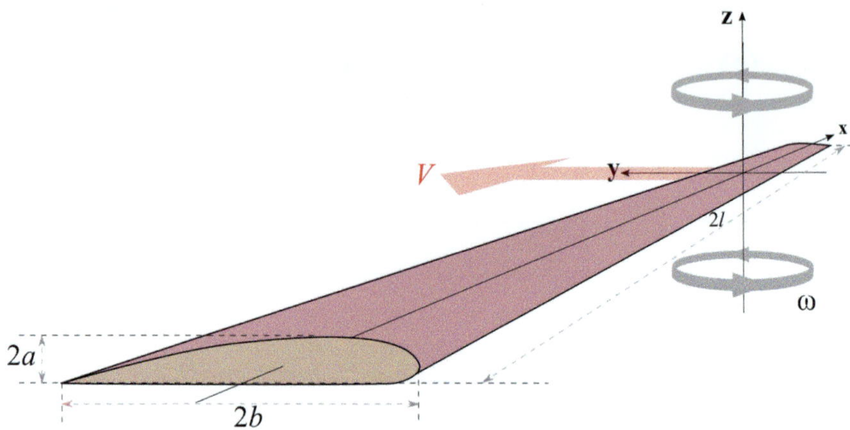

Fig. 7 A bad boomerang—it won't fly. I show it because we will need the notation in TA-3 to explain torque and stability. Here the flight velocity along the **y**-axis is ***V***. Note that the arms rotating about the **y**- and **z**-axes (of length l) are long, whereas the "arms" rotating about the **x**-axis (of length b) are short. This short length is responsible for flight instability. A bent boomerang has a longer b and so is more stable

smaller than l). Now say we start rotating the NRB about each of the axes **x**, **y**, and **z**. How far do the corners move? Clearly if we spin about the **y**- or **z**-axes then they move much further than if we spin about the **x**-axis. Thus it takes significantly less effort to start the NRB spinning about its long axis.

Now we consider Pete throwing his NRB. It initially moves along the **y**-axis with speed V, while rotating about the **z**-axis with angular speed ω (see Fig. 7). Pete has not yet tuned his NRB perfectly, and there remain unwanted, residual, small torques (we label them $G_{x,y,z}$) about all three axes. The torques $G_{y,z}$ are tolerable because, as we have just seen, it is hard for them to get the NRB rotating about these axes. G_x is more of a problem because it will quickly cause a rotation. Thus if $G_{x,y,z}$ are all the same magnitude then in a small time interval δt we will find that G_x causes the NRB to rotate an angle $\delta\phi$ about the **x**-axis whereas $G_{y,z}$ will give rise to much smaller rotations about their axes. The **x**-axis rotation may then lead to deviation of the NRB trajectory from a straight line, as we have seen, due to increased angle of attack, therefore increased lift force, etc. Here is the key to spin stabilization: Pete's NRB is rotating about the **z**-axis right from the get-go, and this limits the angle of rotation about the **x**-axis due to G_x. Why so? After half a cycle of rotation about **z** *the direction of* G_x *has reversed.* The torque G_x causes the plane of the NRB to rotate $\delta\phi$ about the **x**-axis after half a cycle, and then to rotate in the opposite direction back down to zero after the next half cycle. Thus as Pete's NRB flies through the air, this plane flutters over an angular range $\delta\phi$

rather than spins end-over-end without limit as the ruler did. This fluttering will degrade the NRB performance, but the degradation will be much less if $\delta\phi$ is small than if it is large. We have done the math [23] and see that $\delta\phi$ decreases as NRB rotation rate ω increases—it is about 10° for typical ω. $\delta\phi$ is big enough to cause problems for a straight NRB but if the NRB is bent (so that the dimension b of Fig. 7 increases significantly—typically by a factor of three to five) then $\delta\phi$ is reduced by the same factor. So finally we see why no boomerangs are straight—the flutter angle would be too great; keeping it at 1–2° is tolerable, but more than that is uncool.

Finally, how do the unwanted torques $G_{x,y,z}$ come about in the first place? There are many reasons. We have already discussed the main one: different lift forces acting on the lifting arm and dingle arm. Here is a second source of torque which arises, ironically, from the bent shape of boomerangs. For an asymmetric boomerang shape it is possible for the center of mass (the balance point) to lie outside the boomerang and, more importantly here, for it to be different from the center of rotation. (This is not possible for a symmetric boomerang, for which center of mass and center of rotation are necessarily the same point.) If they are not the same, then the velocity of a given point on the NRB surface changes as the NRB rotates, which gives rise to different lifting forces, and so to torques.

7 Summary

Boomerangs are ancient throwing weapons which evolved independently many times around the world, and which developed to different degrees, the epitome being the returning and non-returning boomerangs of Australia, and the rabbitsticks of North America. Returning boomerangs are thrown with the plane of rotation nearly vertical; precession causes them to follow a curved path. Non-returning boomerangs are thrown with the plane of rotation horizontal; they fly straight due to meticulous tuning—altering the surface shape—to eliminate the various torques that deflect the trajectory. A well-tuned NRB can fly straight for much further than a rock or stick can be thrown, and it is easier to aim at a target.

Understanding boomerang flight requires understanding precession and aerodynamics. For this reason boomerang physics is harder than that of other ancient projectile weapons, even while the weapon itself is simpler.

The impressive capabilities of NRBs made them useful hunting weapons over flat and open terrain such as deserts. The fine-tuning requirements matched the weapon to the thrower—a hunter would be better throwing his own NRB than one he did not tune himself. This fact explains why NRBs were not employed as projectile weapons for large-scale warfare: they were difficult to make and required great expertise.

References

1. L.G. Blackman, Aboriginal wooden weapons of Australia: illustrative of the collection of the B.P.B. Museum, vol. 2. Occasional papers of the Bernice Pauahi Bishop Museum, p. 48 (1903)
2. D.S. Davidson, Is the boomerang oriental? J. Am. Oriental Soc. **55**, 163–181 (1935)
3. F. Hess, Boomerangs, aerodynamics and motion. Doctoral thesis, Groningen University, The Netherlands, 1975. http://revedeboomerang.free.fr/Boomerang.pdf
4. J.C. Vassberg, Boomerang flight dynamics, in *30th AIAA Applied Aerodynamics Conference (25–28, New Orleans, LA)*, pp. 2650–2692 (2012). https://doi.org/10.2514/6.2012-2650
5. P. Gudem, Flight dynamics of boomerangs: impact of drag force and drag torque, in *AIAA Aviation 2020 Forum* (AIAA, 2020), pp. 2020–2709
6. R. Perry, G. Perry, https://boomerangs.com/en-ca/pages/how-to-throw-boomerangs
7. See https://www.youtube.com/watch?v=jPiT7RFsfwY
8. R.B. Smyth, *The Aborigines of Victoria*, vol. 1 (Government Printer, Melbourne, 1878)
9. F. Hess, The aerodynamics of boomerangs. Sci. Am. **219**, 124–136 (1968)
10. Dublin University, Dublin University Magazine (1838), pp. 168–171, https://babel.hathitrust.org/cgi/pt?id=mdp.39015030946555&seq=185
11. L. Bordes, A scheme of evolution for throwing sticks. EXARC (2024), https://exarc.net/ark:/88735/10742
12. D. Leder et al., The wooden artefacts from Schöningen's Spear Horizon and their place in human history. PNAS **121**(15), e2320484121 (2024)
13. Dr Philip Jones of the South Australia Museum (which houses the largest collection of boomerangs in the world) is an expert on the subject. His view is that NRBs would be very accurate at 40 yards range, and that the large hunting boomerangs found in Central and Eastern Australia would have achieved ranges of 100 yards. P. Jones 2025, personal communication
14. J. Jennings, N.H. Hardy, The Boomerang and its Flight, vol. 2 (The Wide World Magazine, 1899), p. 627
15. P. Valde-Nowak, A. Nadachowski, M. Wolsan, Upper Paleaolithic boomerang made of a mammoth tusk in south Poland. Nature **329**, 436–438 (1987). Philip Jones, a historian at South Australia Museum, has expressed to me some skepticism concerning this Polish boomerang, pointing out that the shape is rather more

rounded than any Australian boomerang and that the location of the find can be understood as a ritual site

16. K.N. Smith, A mammoth tusk boomerang from Poland is 40,000 years old. Ars Technica, https://arstechnica.com/science/2025/06/a-mammoth-tusk-boomerang-from-poland-is-40000-years-old/. Accessed 30 Jun 2025
17. F. Hess, A returning boomerang from the Iron Age. Antiquity **47**, 303–306 (1973)
18. A. Pitt-Rivers, Opening address to section D, sub-section anthropology of the British association meeting at Brighton, 1872. Nature **6**, 323-4, 341-3 (1872). Lieutenant General Augustus Henry Lane Fox Pitt Rivers was a British soldier, ethnologist and archaeologist who wrote, confusingly, under two names, Lane-Fox and Pitt-Rivers. He was a prolific writer and was clearly fascinated by boomerangs. On the Egyptian boomerang and its affinities. J. Anthropol. Inst. Great Britain **12**, 454–463 (1883)
19. See the National Museum of Australia website, Earliest Evidence of the Boomerang in Australia, https://www.nma.gov.au/defining-moments/resources/earliest-evidence-of-the-boomerang-in-australia
20. See the National Museum of the American Indian website, Throwing Stick/Boomerang, https://americanindian.si.edu/collections-search/object/NMAI_43462
21. L. Bordes, A. Lefort, F. Blondel, A Gaulish throwing stick discovery in Normandy: study and throwing experimentations. EXARC (2015), https://exarc.net/ark:/88735/10203
22. See the Department of Museums, Chennai website, Valari (Boomerang), https://govtmuseumchennai.org/museum-section/ethnology-gallery/valari-boomerang
23. M. Denny, Non-returning boomerangs: an engaging exercise in rotational dynamics. Eur. J. Phys. **46**, 025002 (2025). https://doi.org/10.1088/1361-6404/adb65f

4

Bows and Arrows

Chapter Summary The origins, evolution, and widespread use of historical bows are presented, emphasizing the variation in design (self-bows, composite bows, longbows). Bow construction is briefly summarized. The influence of bow geometry on the internal ballistics is discussed. Arrows need to be matched to the bow, to ensure a smooth launch and flight. The trifecta of Eurasian Steppe geography, horses, and composite bows is introduced. The profound and destructive influence of horse-archer nomadic pasturalists (Huns, Mongols) on Eurasian history is emphasized. A simple model of bow physics explains many observed features of bow dynamics: draw weight, arrow launch speed, efficiency, performance differences between types of bow. English and Japanese longbows, and their influence upon history, are contrasted. The Archer's Paradox is discussed and resolved.

"From about 700 B.C. to A.D. 500, the vast territory of Scythia, stretching from the Black Sea to China, was home to diverse but culturally related nomads. Known as Scythians to Greeks, Saka to the Persians, and Xiongnu to the Chinese, the steppe tribes were masters of horses and archery."—(Adrienne Mayor, American Historian) [1].

1 Projecting Power: Introducing the Bow and Arrow

Probably the first weapon developed by humans that could potentially kill a person 200 yards away, or kill the archer's dinner at that range, the bow was used on all occupied continents except Australia from prehistory to the gunpowder

M. Denny, *Slings and Arrows*,
https://doi.org/10.1007/978-3-032-08563-4_4

era. Bows were widely adopted across Eurasia from the 2nd millenium BCE as *en masse* weapons of war, arrayed in serried ranks or as looser lines of skirmishers in front of enemy formations. In these early centuries they were also used widely by charioteers. By way of introducing these weapons to you, we need to discuss a few bow basics before delving into the historical/technological details.

1.1 The Bow...

First: bow anatomy and terminology. The *belly* is that part of the bow facing the archer when he fires; the *back* is the part facing the target. These parts are differentiated because they are subjected to different forces when the bow is *drawn*—when the bowstring is pulled back. The belly is compressed whereas the back is stretched. *Brace height* is the distance between the middle of the bowstring and the grip, when the bow is braced. *Draw length* is the distance that the bowstring is pulled back, plus the brace height. Loosely, it is the distance between the archer's hands when the bow is drawn. (See the box ***Power stroke and draw length***.) The *arrow shelf* is a groove cut in some bows, just above the handle, where the arrow rests. (Some simple bows had no shelf; the arrow rested on the archer's hand.) Finally *draw weight* is the force needed to draw the bow.[1]

The first bows were simple *self-bows* meaning that they were formed from a single stave of wood (ash, elm, and yew were favored). The more sophisticated (in terms of manufacturing) *composite bows* arrived surprisingly early in history. These were made from a number of different components, glued together to form a more powerful weapon than a self-bow of the same size. Either type could be a *recurved* or more extreme *reflex bow*, making the bow more powerful when strung—we will delve into details later. There was an intermediate type, possibly a precursor to the composite bow: the *two-wood* bow was common in northern Eurasia. It consisted of a pine belly and birch back, and was widely used in boreal regions from prehistory well into the second millennium [2].

Most archers hold the bow in a vertical plane, or nearly so, when firing. This was true in the past as much as it is true today. A few historical archers, notably native North Americans, favored a horizontal stance. There are pros and cons to each, and we will deal with these here and now because it is a really basic aspect of the weapon. The horizontal stance is more convenient when shooting from a crouched position, as when hunting an animal or ambushing a human enemy. It is more convenient when on horseback—the archer can swing their

---

[1] BTW we say that the archer is right-handed if he draws the bowstring with his right hand. Sometimes the bow itself is called right- or left-handed, depending on which side of the grip the arrow shelf is located.

bow left or right without knocking the horse's head. It is more convenient for shooting a very large bow (if the distance from an archer's shoulder to the ground is D, and the length of a symmetrical bow exceeds $2D$, then the archer cannot hold the bow vertically while firing the arrow horizontally).

Power stroke and draw length

The *power stroke* is another technical term that appears in archery literature and online forums. It is the distance over which the arrow is accelerated, or, in other words, it is the distance that the arrow is pulled back when fully drawn. For me, *this* should be called the "draw length," because it is the length that the bowstring is drawn. The physicist in me doesn't like the term "power stroke" because it is misleading: it is not a stroke and it doesn't have units of power.

If a bow is a *straight bow*, i.e., one that is straight when strung, then draw length and power stroke are the same (more or less—experts would quibble over the thickness of the bow handle) because the brace height is zero for straight bows.

In this book I will use the term "draw length" for both true draw length and power stroke—the difference is small for most bows. A reader who is bothered by the resultant ambiguity will likely be expert enough to tell the difference from context.

These advantages are usually outweighed by the disadvantages. First and foremost, it is more difficult to aim a bow held horizontally. Imagine that you are a Neolithic hunter eyeing a deer to feed your family: you may get only one shot at your target. Holding your bow vertically allows you to bring one eye close to the line of the arrow, making it easier to aim—it is just anatomically more awkward to sight an arrow when the bow is horizontal. Second: in a military formation, massed ranks of archers can be placed closer together if they hold their bows vertically than if the bows are horizontal, and so the firepower that can be brought to bear on an enemy formation is greater. Third, a bow held horizontally cannot be drawn as much as the same bow held vertically—again, this is simply a matter of geometry and anatomy. English longbowman firing at an advancing formation of French heavy cavalry during the Hundred Years' War needed maximum range—an acre of cavalry does not need much accuracy to hit, and you want to hit them early—and so their first salvo would be at long range requiring maximum draw.

So, most archers held their weapon nearly vertical, even those history-changing horse-archers we will describe in the next section. As we will soon see, early on some of them made design changes to their bows so that they could be held vertical on horseback rather than hold their bows in the more convenient horizontal stance, such were the overall advantages.

Another basic feature of archery that we need to address here and now is the grip. In the western part of Eurasia the bowstring was drawn by the tips of two or three fingers. Usually these fingers are immediately below the arrow, but for the *split-finger* grip the index finger is above the arrow. For such a grip the arrow shelf is on the left side of the bow handle (for a right-handed archer, confusingly). In North America the *pinch* grip was favored—the fingers hold the arrow and not the bowstring. In the Far East, it was common for archers to draw their bows with their thumb, not fingers. Usually they wore a metal *thumb ring* to avoid strain or injury. For the thumb draw, the arrow shelf is on the right side of the bow handle, for a right-handed archer. Both grips worked; they may have required different drawing techniques but yielded similar results.

1.2 ...and the Arrow

Terminology: see Fig. 1. The *arrowhead* is the pointy bit, made of stone in the distant past and then metal. The *nock* is the groove at the other end of the arrow that holds the string when the bow is drawn. The *shaft* is the—no surprises here–shaft, or body of the arrow. Its stiffness is an important characteristic, and is called the arrow's *spine*. The *fletching* is the feather vanes at the back, needed for aerodynamic stability.[2]

An arrow's length had to exceed the bow's draw length, of course, but not by much. We will see at the end of this chapter that there were and are more basic connections between a bow and the arrow it fires; arrows are matched to a bow in just the same basic way that bullets are matched to a gun. An arrow's spine must be just right, so that the missile can slide past the bow grip without veering off course. This behavior is part of the infamous/mysterious (if you are into archery/not into archery) *archer's paradox*. Intrigued? Read on.

2 Bow Structure and Evolution

This chapter will continue the pattern of earlier chapters by blending history and science/technology. (Algebra and atlatls, ballistics and boomerangs, trigonometry and trebuchets—you get the idea.) The aim is to explain the

[2] The surname Fletcher comes from the trade of making arrows, as does Arrowsmith. The surname Bowyer comes from the trade of making bows and, of course, Archer from using them. A Stringer made bowstrings. The large lexicon of archery terms, and their creeping into people's names, points to the historical importance of archery in England. Had this book been written in Korean or Turkish or Hunnic, we would find the same linguistic adoption of archery words (e.g., Jang, Okan, ?). Bows mattered a lot, in the past.

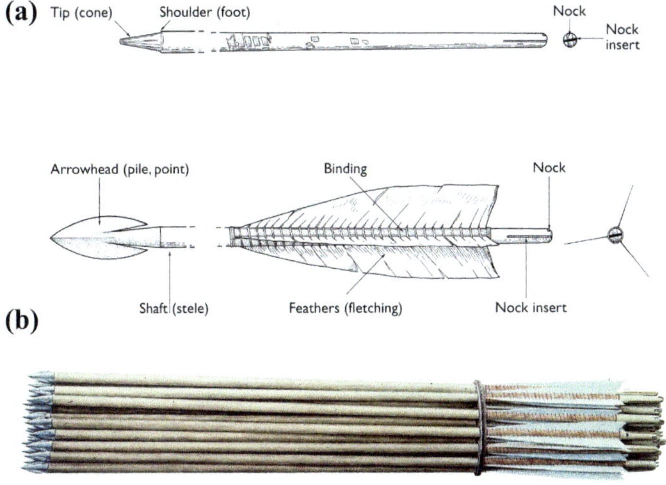

Fig. 1 Arrows. **a** Arrow anatomy. **b** A few of the 2,000 arrows recovered from the *Mary Rose*, an English carrack of Henry VIII's navy which sank in 1545. ©Mary Rose Trust

physics that underpins bows and arrows (and all our other ancient projectile weapons) and to show where they fit into history and how they have profoundly shaped history. There is no better place to begin such an account about archery, in order to illustrate the interaction between history and technology, than with the development of bows. The reason is this: bows represent the most complex technology that we have thus far encountered in this book, by a country mile, and the story of their development has more, and longer, strands to it than any other of our ancient projectile weapons. (And so this is the longest chapter of the book—until we get to siege engines.)

No doubt the very first bows were short self-bows—basically a stick, bent and braced with a sinew or fiber bowstring. These bows are the simplest to make. In the following is a basic and crucial fact: small bows are less powerful than big ones. Even if the draw weight is the same, the small bow will be less powerful (see box ***Scaling bows***). Bows started out in the Paleolithic as relatively weak projectile weapons. Later bow developments—described herein—led to short bows with increased draw weight and to larger longbows, both of which generated higher arrow launch velocities and so both longer ranges and better armor- (or deer- or pig-) penetration.

Scaling bows

When we get to the analysis of bow physics later in this chapter we will see that the energy of an arrow that has just been released is approximately proportional to the draw weight multiplied by draw length. Consider now two simple self-bows, one of which is bigger than the other by a scale factor S. That is to say, every dimension of the larger bow is bigger than that of the smaller bow by this amount: bow length, draw length, and bow thickness. Empirically we know that the stiffness of a bow—its resistance to bending—increases geometrically[a] with limb thickness as $S^{1.6}$. We see that, comparing a 6-ft self-bow with an identical (except for scale) 4-ft self-bow, the scale factor is $S = 1.5$. Thus our knowledge of how bows scale tells us that the energy of an arrow fired from the bigger bow exceeds that fired from the smaller bow by a factor of approximately 2.9. Bigger bows have much more killing power, all else being equal.

[a] Clifford Rogers, personal communication, February 17, 2025. Rogers is Professor of History at the United States Military Academy at West Point, an expert at medieval European warfare in general and English longbows in particular.

A good example of a short self-bow is that used by Native Americans. The Comanche bow was made of a single stave of wood, often osage orange, of length $3\frac{1}{2}$–5 feet, and with a draw weight in the range 40–50 lb. This would be a short-range bow, often fired from horseback. The Comanche were an aggressive nation and were frequently at war with their neighbors, be they other native Americans or European colonists. They were well known for the rapid rate of fire that they could achieve with their bows. It is likely that these bows were brought across Beringia (the land mass connecting East Asia with Alaska during the last Ice Age, when sea levels were low), perhaps several times. Given the low population of North America in antiquity, compared with that of Eurasia, it seems less likely that local innovations led to any radical changes in bow design, as happened in the Old World.[3] The evolution of bows in North America thus has probably been more influenced by new types of bow coming across from the Old World than by local developments [5].

These early bows were symmetric, or as nearly so as a tree branch will permit. By symmetry we mean simply that the shape of a strung bow to the left of the grip is a mirror image of the shape to the right. Place the middle of the grip on your extended finger: the bow balances horizontally. Another example of a symmetric self-bow is the English longbow of the Middle Ages, which we will

[3] Generally it seems that the rate of technological innovation increases with population size, at least for low populations [3, 4].

discuss later alongside the unusual asymmetric Japanese longbow. The size and draw weights of a number of historical bows are shown in Fig. 2.

The simplest self-bow is a bent stick, and its power comes from the native stiffness of the wooden stave that forms it. Some types of wood are better than others for making self-bows (such as hickory and hard maple in North America, yew and ash in Europe, bamboo and mulberry in East Asia) and arrows (such as southern arrowwood—appropriately—hazel, and bamboo). For ancient bows, a branch of the right size would be selected and string grooves cut into the ends to hold a bowstring (sometimes an *eye*—a small hole—would be made at each end for attaching the string). Perhaps an arrow shelf would be cut, above the handle. Historically bowstrings were made from silk (in Asia), sinew, and from hemp, nettles, or other plant fibers—more on bowstrings later.

As the centuries went by, bowyers learned how to improve their product. By the Middle Ages in Europe, bow staves would be cut from the main trunk of trees (along their length so that the grain ran along the bow, not across it, for strength), such that the strong heartwood formed the belly and the springy sapwood formed the back. The cross section of the bow would vary along its length, being thicker near the center and flatter toward the ends. Draw weight increased from maybe 25–80 lb in the Neolithic to 80–110 lb in medieval times. (See Fig. 2: these numbers are typical, but draw weights outside these ranges certainly existed—some English longbows had draw weights of 180 lb [6].) Arrowheads changed from flint in the Stone Ages to metal, with different shapes for different purposes (with broadheads for hunting, and bodkins for armor penetration).

Short bows became more powerful when hunters and warriors moved on from the simple self-bows to *composite* bows. These more sophisticated weapons developed surprisingly early, given their complicated manufacture: it is thought that composite bows originated on the Eurasian Steppes[4] about 4,000 years ago, and spread from there across the Old World. As for their manufacture, the stages are fairly uniform, with interesting regional variations. Perhaps this fact suggests to you that composite short bows were invented only once, and proved so successful that they were quickly adopted all along the Steppes and then further afield. Or maybe they were invented several times independently, and the uniformity of construction reflected only the limited materials available four millennia ago. However it arose, the composite short bow was stronger than a self-bow of the same size. Stated differently, given two short bows of comparable performance, the composite would be significantly smaller—very desirable for mounted archery, as we will see.

[4] Mid-latitude grasslands that stretch 5,000 miles east-west, the Steppes were and still are ideal terrain and pasture for horses.

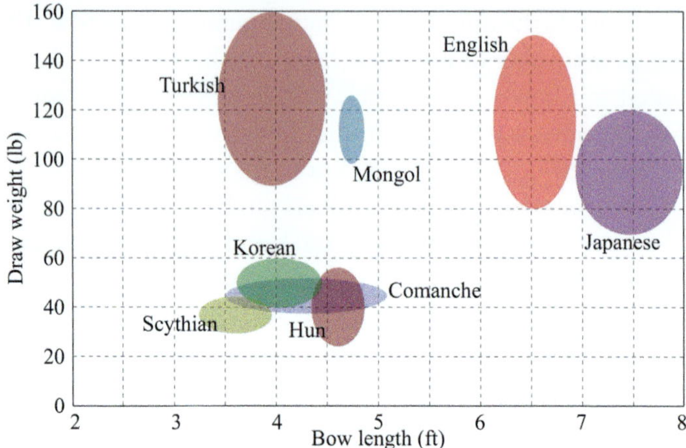

Fig. 2 Draw weight *versus* length, for a number of bows from classical antiquity and the Middle Ages, say 500 BCE to 1500 CE. The English and Japanese bows are longbows; the others are short. The English and Comanche are symmetrical self-bows; the others are composite and asymmetrical. The Scythian and Hun are from antiquity; the others are medieval or later (these data are for typical bows: outliers exist for all of them)

So, how were composite short bows made? Most were a laminate of horn (thin strips, on the belly side), wood (core), and sinew (on the back). These components were glued in place with a hide glue or fish bladder glue. These glues were strong when set, though susceptible to weakening when wet. It is no coincidence that self-bows dominated in the wetter parts of the Old World—in such regions the composite bows needed to be protected from the weather or else carefully lacquered. The horn was from water buffalo or ibex, usually—it seems that cattle horn bows didn't last long. The middle of the sandwich consisted of the wood (again, not just any wood—there were preferred species and preferred parts of the tree, as we have seen), sometimes steam-bent, which defined the basic shape of the bow. The sinew was usually from the lower legs of deer, horses, or pigs. After being glued in place—with the fibers running lengthwise down the bow—the sinew would shrink considerably causing the bow to recurve, or reflex to the point where the bow ends touched. Unbraced, braced, and drawn bows are sketched in Fig. 3. The bowstring of a recurved bow was under greater tension than that of a self-bow, and so the draw took more power right from the beginning of the pull-back. The greater draw weight of these short composite bows arises in part because the horn belly has greater compressive strength than wood, and the sinew back has greater tensile strength.

The draw characteristics of recurve bows are different from that of bows that are straight or weakly recurved. For two bows (one recurved and one not) with

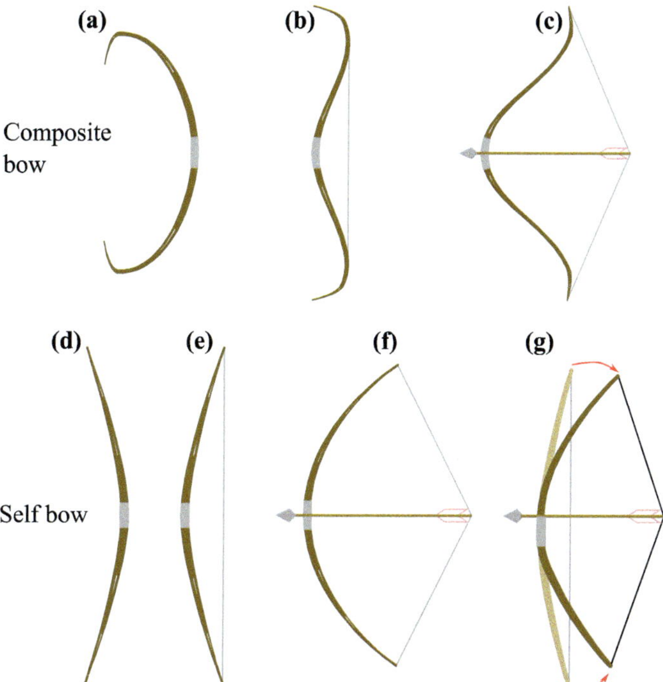

Fig. 3 Bow profiles. Symmetrical recurved composite bow (**a**) unbraced, (**b**) braced, and (**c**) drawn. Note the difference in length of the bowstring (the section that is not touching the shaft). Symmetrical self-bow (**d**) unbraced, (**e**) braced, (**f**) and drawn. **g** In reality all bows are asymmetrical to some extent, due to the grip and the arrow shelf. The shorter and thicker lower limb is displaced less than the upper limb when the bow is drawn

the same full draw weight, the recurve will have a higher initial draw which may ease off as the draw approaches the full draw length. A non-recurved bow usually has a more linear draw weight (i.e., the force increase with draw weight is pretty much the same at the beginning and end of the draw). Composite bows were always recurved, sometimes very strongly whereas self-bows were not, and so the two bows feel different to archers because of these different draw characteristics.

Over the centuries, improvements arose. An early improvement, which we will discuss when we get to the Huns, is an asymmetric profile. (There is some asymmetry in every ancient bow—see the box ***Tiller***.) A second upgrade, made in the fourth-century BCE, is the *siyah*, an Arabic word describing stiffened (usually horn) "ears" added to the ends of the bow. These increase draw weight but also reduce vibration.

Tiller

The grip of a perfectly symmetrical self-bow is placed right in the middle of the shaft, of course. But the fact that this part of the bow is held by the archer introduces a small but significant asymmetry. Let us say that there is no arrow shelf carved out of the bow (which would introduce a left-right asymmetry, as the archer views the bow). In that case the arrow rests on the archer's hand and so is not exactly half way between top and bottom ears (edges) of the bow—the symmetry is broken. More often the arrow is right in the middle, in which case the grip must be a little below center. In either case there is a slight difference in length of the flexible part of the upper and lower limbs. This difference has the consequence that the shorter limb moves more when the bow is drawn, as shown in Fig. 3g. But the limbs must return to the braced position at the same time when the arrow is released, if it is to fly straight, and so the shorter limb must be stiffer (made so, for example, by virtue of a slightly thicker shaft, as suggested in Fig. 3g). Bowyers must take this asymmetry into account when manufacturing their bows; *tiller* is the name they give to a measure of the limb asymmetry.

2.1 The Classical Trifecta

In military terms, three factors came together in the fourth millenium BCE that changed warfare and history, then and at many later times. The three components that led directly to the nomadic horse-archer cultures of classical antiquity (Scythians—see Fig. 4, Sarmatians, Parthians, and Huns, among others) and the Middle Ages (the Mongols) are these: The Steppes, the horse, and the composite bow. I call these a trifecta, though "perfect storm" might be a better characterization.

Horses were likely first domesticated somewhere on the Eurasian Steppes in the fourth millennium BCE. Their use spread rapidly during the third millenium BCE, and the size of individual animals grew significantly [7]. This growth, due to selective breeding, continued from prehistory into the historical era [8]. The reason for this selective breeding is obvious: larger horses can pull heavier wagons or plows. In terms of warfare, small horses could pull chariots, but cavalry had to wait for the development of horses that were big enough to carry a warrior.

Of all the common domesticated creatures, the horse is the pickiest eater, in the sense that it is the most demanding of pasture land: it needs lots of grass. Cattle are less fussy—they can be put out to pasture on poorer land than

Fig. 4 Detail from a fourth-century BCE electrum beaker found in the Kul-Oba kergun (barrow—burial mound) in the Eastern Crimea. It shows a Scythian warrior bracing his composite bow. Scythian bows were symmetric, and small enough to be handy for these nomadic horse-archers. (Public domain: author unknown)

that required for horses. Sheep are even less fussy—there are many parts of the world where sheep have been introduced that are unsuitable for cattle. Goats will get by in places where sheep could not and, lastly, camels live in deserts. Meaning, our domesticated dromedaries and befriended bactrians can live on thin air, or at least they can find forage on land where none of the others can live [9]. (Chickens—the other common domesticated animal—probably fit closer to the camel end of the spectrum than to the horse end, in terms of required habitat.)

Of these animals, only three can be utilized for transport: cattle—usually oxen—are good for hauling carts and wagons, but not for riding or for pulling chariots. Other quadrupeds (zebra, llama, pigs...) were unsuitable for one reason or another—a pity, because a zebra cavalry unit would have looked cool, and I would pay good money to see a troop of pig-mounted cataphracts or hussars. Only horses and camels have the physical ability and temperament that allow humans to train and ride them.[5] Camels are limited in terms of terrain and climate: they prefer arid regions. Horses can be ridden over a much wider range of terrain and climate, and they are generally faster and more maneuverable than camels. So, horses have been the animal of choice for people in

[5] Thus, there were no camels and no domesticated horses in the New World, pre-contact, and so no cavalry.

antiquity to get from A to B as conveniently as possible. And they were made for the Steppes.

Of course, the phrase "getting from A to B" is not meant to imply horse rider literacy. It is no coincidence that horses were domesticated on the Eurasian Steppes (thousands of square miles of pasturelend) and it is a sad fact (for historians) that two, three, four millennia ago these lands were populated by mostly illiterate people. Not writing things down means that we do not have their versions of events. However, in classical antiquity this vast region—western and central Asia—was bookended by two literate cultures. Rome and Han China had little direct contact with each other[6] and they both had problems dealing with the horseriding nomads of the Steppes. They both wrote about "barbarian" nomads, who they fought against many a time, and so this is how we know a little about Steppe people of this period—biased views of their enemies (plus—increasingly—unbiased but incomplete results from archaeology). All of these nomadic people were horsemen (and horsewomen and horse children). Many, but not all, in the classical era conducted warfare as light horse archers, meaning that the warrior and his horse were unburdened with heavy armor, and carried a short composite bow as their principal weapon. The nomadic pastoralists' strategies in war and their tactics in battles were both predicated upon the skills of their horsemen as riders and archers.

So the trifecta formed. Very skillful horse archers armed with very good bows and riding the best horses swept across thousands of miles of easy terrain with plenty of pasture. These horse archers have had a significant impact upon the history of the Old World (a devastating impact for several of the "more advanced" civilizations they encountered); they formed the basic unit of armies that scoured the Old World in both the first and second millennia CE, and it is now time for us to take a look at two groups of them—from a safe distance.

2.2 Huns

A Hun horse archer and his horse were both tough as boot-nails. (No doubt their wives and kids were tough as boot-nails, too.) Huns were a surprisingly diverse group of people, about whom we have learned very little. They did not leave us any written records telling us about themselves and so most of what we know comes from their enemies. Historians of antiquity knew nothing about the origin of the Huns; 200 years ago their successors thought they had figured

[6] The first direct contact occurred in 166CE when Roman Emperor Marcus Aurelius sent an envoy to the Han Court, [10] so we know for sure that these two giants of the ancient world knew of each other, but most contact was indirect through trade along the Silk Road.

out where the Huns came from, and today historians are again unsure—all this uncertainty arises from the lack of Hunnic written records.

Han China fought barbarians on their borders, as did the Romans (by "barbarians" both meant, of course, enemies who were not Chinese, or Roman). One such group—nomadic horse archers—were the *Xiongnu* (or *Hsiung-Nu*), a large and loose confederation of raiders. We know from imperial sources that the Han armies forced a division of these people into two groups, the northern and southern Xiongnu, in 48CE. The southern group were defeated,[7] and then absorbed into Han China. The northern group headed westward for new pastures, and were last mentioned in Chinese records in 93CE. This northern group have been identified by many historians as the Huns who, nearly three centuries later, burst out of the Steppes to torment Europeans, both Roman and non-Roman. Some scholars regard the names "Xiongnu" and "Hun" as cognates, but there is no consensus among the historical community because we are not even sure what linguistic group the Hunnic language belonged to, let alone what they called themselves.

What little we do know about the Huns comes from the written records of the book-end civilizations of Rome in the West and Han China in the East. For the historian much of the lands in between—the Eurasian Steppes—formed something like a black box in the classical era, about which very little is known because very little was written down by the populations (quite large populations—millions—we now think) who lived there. At the end of the first-century CE the Xiongnu headed west into this black box, and in the 370s the Huns headed west out of it. The spatial separation is some 2000 miles. Our scanty knowledge suggests that there are similarities between the Hun and the Xiongnu (both nomadic, with armies built around horse archers, with similar weapons and metal cooking pots) and differences between them (political organization, hair styles). The similarities may be attributed to coincidence, or the differences may be attributed to three centuries of cultural evolution, as you see fit. There is no consensus among scholars, though most think there is at least a connection between Hun and Xiongnu.

Whatever doubts that we have about their origins, there is none about their fighting ability. Huns were fierce warriors, and were universally regarded as such by those people in the West who fought against them. "*A race savage beyond all parallel*" is the well-known and oft-repeated description (due originally to Ammianus Marcellinus, a fourth-century Roman historian[8]). Savage in

[7] Defeated by powerful Chinese cavalry consisting of their prized "heavenly horses" bred in the Ferghana Valley of modern Uzbekistan.

[8] Marcellinus also said of the Huns: "*This active and indomitable race, being excited by an unrestrained desire of plundering the possessions of others, went on ravaging and slaughtering all the nations in their neighborhood till they reached the Alani. . .*" These Alans, from the Pontic Steppes, later became Hun allies.

warfare, probably; a race, not so much. Their armies were very successful: within a few decades of arriving in eastern Europe they had conquered most of it, though conquered in the sense of defeating its defenders and despoiling their cities—it seems that Huns were interested in booty more than territory (they were pastoralist, not agrarian). Likely there was no grand plan of conquest with a single emperor directing traffic; the (in)famous Attila being an exception. Usually Hunnic armies were a mixture of different people, unified and working together well on the battlefield, but much less cohesive politically. Ethnically they were a mixture of widely different groups. By the time they were rampaging through Central Europe they had Germanic and Iranian allies, sometimes arrayed separately as distinct units on the battlefield and sometimes integrated culturally and socially—despite their (probable) East Asian ancestry the Huns seem to have been racially and ethnically mixed. They incited fear in opposing armies—those of Goths and other Germans, and of Romans from both Eastern and Western Empires. This fear was due to their tactics on the battlefield (very rapid and coordinated movements, feints, fearlessness, bowmanship—of which more shortly) and to their physical appearance. Roman writers considered them, their clothes, and even their horses all to be ugly. It seems that Huns scarred their faces, and some of them practiced cranial deformation, which was not pleasing to the eye of their Western enemies.

And then there was the "golden bow," a valued symbol of identity to the Hun and of power to the Hunnic nobility. It was a short, composite bow of the type we have described, but differing from the original in one crucial respect: it was asymmetric. The upper limb was noticeably longer than the lower (perhaps 30 inches *versus* 24), which gave the mounted archers an advantage—they could switch the bow easily from side to side without getting snagged up with their steeds' necks. Most horse-archers used short, asymmetric bows for this reason. Another Hunnic innovation[9] was to add bone plates to the siyahs and to the belly of the riser (that part of the bow center either side of the grip). These additions made the Hun bow sturdier, but heavier. So valued was their bow that it was often a feature of Hun burials that have been excavated by archaeologists—Hun leaders wanted to be buried with their bows. Their tactics in battle were typical of Steppe nomads, but were executed better than most. Horse-archers would ride fast up to and past enemy formations, loosing arrows all the time and then abruptly retreat, firing arrows behind them (firing

[9] Or possibly the Huns may have taken the idea of bone plates from the Xiongnu—another possible connection. One innovation that the Huns did bring with them from the East was an early form of stirrup, invented in China around 2000 years ago, and providing horsemen with a fighting platform that enhanced their capabilities. Another was the thumb ring for drawing bows.

behind from a moving horse[10] took skill in both archery and horsemanship). Rinse and repeat. Then, when enemy numbers had been sufficiently reduced, the Huns would move in to finish them off with their swords.

Hun strategy was less well organized and coherent than their tactics. They barreled westward, pulverizing any other people in their path, be they settled agrarians or nomadic pasturalists. The Romans had heard of these barbarians from the East in the second-century CE, but did not cross their path until the fourth. Hundreds of thousands of displaced Goths—whole communities, not just defeated armies—fled west and south to take refuge inside Roman territory. Sometimes a deal was made, and the Goths would sign up for service in the Roman army in return for sanctuary for their families. Sometimes they burst the gates and forced their way in. These Goths and other people (Sarmatians, Alans, Burgundians, and others) from the Pontic Steppes or Western Asia or Eastern Europe had been pushed westward by hordes of Huns. "*A sagittis Hunnorum nos defenda domine*"—From the arrows of the Huns defend us O Lord.[11]

Their bows and horses took the Huns under Attila (Fig. 5) as far west as France, where they were beaten in 451 CE in the Battle of the Catalaunian Fields by an alliance of Romans (it was one of the Western Empire's last victories) and Visigoths. They retreated eastward and were beaten again by a Germanic alliance at the Battle of Nedau, in 454 CE after Attila's death. They then gradually disappeared from history, likely absorbed into other tribes such as the Bulgars of the Pontic-Caspian Steppe. It is tempting to conclude that the Huns had overreached: Europe west of the Hungarian Plains had no large pasturelands for their horses; the wet weather compromised their bows; the horse-archer mode of fighting was less well suited to the western hilly and forested topography.

2.3 Mongols

Fast forward 750 years and we find, emerging from the Eurasian Steppes, another group of fierce nomadic horse-archers. The Mongols did what the Huns tried to do—pulverize the rest of the world—and they did it better (if that is the right word for meting out mayhem, massacre and misery) over a wider area and for a longer time. Genghis Khan had united the bickering

[10] This is the *Parthian shot* or parting shot, named after the Persian Dynasty—also famous in their day for their light cavalry—that competed with the Romans in the Middle East and Anatolia for two centuries.

[11] Not a contemporary reference, but a Christian prayer made fully 1500 years later, [11] which speaks to the lasting terror inspired by these invaders from the East. (Germans in WW1 were called "Huns" disparagingly by the French and British.) See also https://orthodoxpatrick.ie/saint-patricks-generation/.

Fig. 5 Bronze medal after a fifteenth-century original. *Attila, fléau de Dieu*—"Attila, scourge of God." Public Domain

Mongol tribes in the first years of the thirteenth-century CE and led them out south, and especially west, to conquer. He got as far as Persia. After his death, successors continued his invasion of the world, and succeeded in grabbing a significant chunk of it: Kublai Khan conquered China. At its peak in the fourteenth century, the Mongol Empire was the largest that the world had ever seen, and the largest contiguous land mass that it would ever see, stretching from Korea to Eastern Europe—some nine million square miles.

Here is a nice quote that I came across when researching the Mongol achievements, which neatly emphasizes the nomadic, Central Asian horse culture that they epitomized: "*Fuelled by grass, the Mongol empire could be described as solar-powered; it was an empire of the land. Later empires, such as the British, moved by ship and were wind-powered, empires of the sea. The American empire, if it is an empire, runs on oil and is an empire of the air.*" [12] As with the Huns, the Mongols spread west very fast along the Eurasian Steppes superhighway (Fig. 6).

Fig. 6 Mounted warriors pursue enemies. Illustration of Rashid-ad-Din's Gami' at-tawarih. Tabriz (?), 1st quarter of fourteenth century. Water colors on paper

The Mongol hordes inspired terror among the people within their newly conquered territories, and fear among those without. The fact is that the Mongols were brutal and merciless beyond the standards or practices of the rest of humanity at that time. The very modern word *genocide* has been applied by more than one historian, to describe Mongol actions [13, 14]. Have you heard of the Tangut Empire (a Chinese dynasty)? No? Very likely this is because Genghis Khan had almost the entire population of three million people killed, written records destroyed, and buildings pulled down. His armies killed between three-quarters and nine-tenths of the population of Persia (the Khwarazmian Empire)—figures are uncertain. They destroyed many ancient cities that offered resistance: Kiev, Moscow, Balkh, Lahore, Baghdad, and scores of others. Cities across territories that lie in modern-day China, Korea, Iran, Iraq, Central Asia, Siberia, the Caucasus, Belarus, Ukraine, and most of Russia either surrendered and paid tribute immediately, or were besieged and then obliterated. Estimates vary widely—perhaps 20–60 million people were killed directly by the Mongol armies. Many millions of others were displaced, or conscripted into military service, or died of the plague (known in Europe as the Black Death) which spread across the Old World during these times, aided by the displacements caused by the Mongol invasions.[12]

[12] It is worth noting that the Mongols practiced *kharash* meaning that they often used captured enemy civilians or defeated soldiers as human shields in their next battle.

Today, Mongolia is very proud of Genghis Khan—he is widely revered. A 40-meter high (130 ft) stainless steel statue of him dominates the Steppe just outside the capital Ulaanbaatar.

Like the Huns, the Mongols were an illiterate people (at the time of Genghis Khan—his grandson Kublai Khan introduced writing across the Empire), who emerged from the Steppe on horseback, with the bow as their principal weapon. Why were the Mongols so much more successful?[15] For sure the Huns, especially under Attila, generated terror and destruction, but relatively briefly and over a relatively small region, say from Western Asia to Central Europe. There are a number of reasons. For one, the size of the Mongol armies was an order of magnitude larger than those of the Huns (hundreds of thousands *versus* tens of thousands). Second: the Mongols were militarily more effective. Their generals were frequently better than those of their enemies (not true of the Huns) and their reconnaissance and communications were outstanding [16]. Their horses were probably better than those of the Huns, and each warrior had three or four, so that he could switch mounts and carry on—Mongol armies covered large distances even faster than did those of the Huns. They had a paired stirrup that permitted a rider to stand up. Lastly, and most importantly to us, these horse-archers had superior bows [17].

Before turning to those bows, a final paragraph on the fate of the Mongol Empire. Though impressively effective and very brutal, the Mongol armies were not unbeatable. The Mamluks (a historical oddity, they were an army of manumitted Turkish slaves that ruled Egypt and Syria in the Middle Ages) beat them twice, halting the Mongol advance in the Middle East. The Delhi Sultanate successfully repelled Mongol invasions. Another Mongol army was ambushed in Russia. In Central and Eastern Europe some besieged cities held out. Mostly, though, the Mongols were beaten by internal divisions. Squabbles over succession, when a leader with absolute power dies, are common in history. The Mongol Empire was divided into four khanates following Genghis Khan's death: the Yuan Dynasty in China, the Ilkhanate in Persia, the Chagatai Khanate in Central Asia, and the Golden Horde in Eastern Europe [18]. These fell apart at different times—first the Ilkhanate in 1335 and then the Yuan Dynasty in 1368. The Golden Horde lasted until the beginning of the sixteenth century and the Chagatai Khanate until the middle of the eighteenth.

What of the Mongol bow, the primary weapon of their military success? It was a better composite bow than any earlier bows, and at least as good as many that came after. Perhaps only the ottoman *hilal kuram* ("crescent moon") bow of the late eighteenth and nineteenth centuries was better, and this was not a war bow.[13] The Mongol bow was shorter than the Hun bow (typically 52 inches versus 55 inches), with no bone stiffeners. No bone component made it less robust, but the reduced weight—particularly in the outer limbs—made it more efficient, as we will soon see. Many Mongol bows had a birch bark "wrapper," or outer layer, for waterproofing. They had a more complex wood core structure and were more recurved than the Hun bow. It was thus more powerful (said to have a range of 350 yards) and easier to use on horseback.

Today many people in Mongolia are keen on archery (with popular horse-archery competitions), as we will see in Chap. 8, but the bows they use are not like those of Genghis Khan's armies—they are much larger and more closely resemble the Manchu Chinese bows that were adopted in the seventeenth century.

3 Bow Forces and Bow Efficiency

Here we investigate the physics, the bow internal dynamics, to see why they are the way they are. We begin with a very simple model of bow draw force which will prove useful for understanding the forces that act, and then adopt a more complex physical model to investigate bow energy efficiency [20, 21].

3.1 Forces

A simple (too simple to be realistic, but not so simple as to be misleading) and instructive model of a bow will serve to get the ball rolling. Consider Fig. 7.

First, we look into the forces that act on a bow when it is drawn. In Fig. 7, we see a symmetric bow stave that is more (a) or less (b) recurved when not braced. These two bows are taken to be the same length and, when braced (c), have the same brace height b. The bowstring is drawn (d) to the full draw weight $F_{1,2}$ over the draw length x_0. Here is my simplification: the force needed to draw the bowstring a distance x is here taken to be constant throughout the pull. The force taken to pull back the bowstring a distance x is then $F = kx$, for some constant k. This assumption is taken to hold also when the bow is being braced so that, for example, it takes a force kb_1 to straighten the recurved bow of Fig. 7a and a force $k(x_0 + b_1)$ to brace it.[14]

[13] The hilal kuram was a specialized flight bow, designed to achieve maximum range, with specialized arrows. Historically ranges of half a mile are claimed, though the modern record is "only" a third of a mile. These bows are not weapons and so are of only tangential interest to us.

[14] For the expert, x_0 is the draw length; the power stroke is $x_0 - b$. Power is transferred to the arrow over this distance, but the draw weight is generated over the distance $x_0 + b_1$.

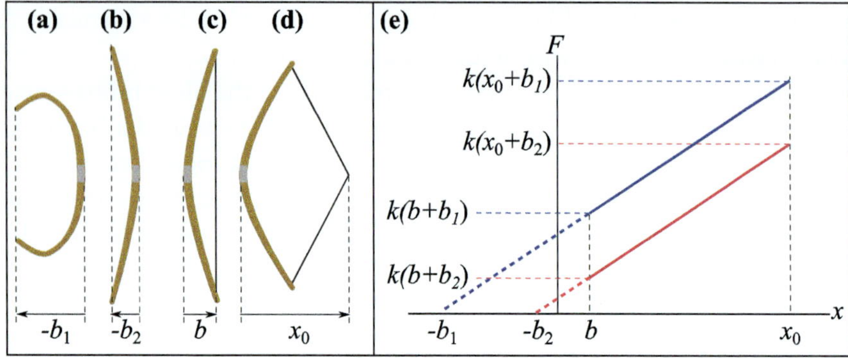

Fig. 7 Very simple model of draw force. **a** For an unbraced recurved bow we can consider the draw distance to be $-b_1$. **b** For a self-bow of the same length it is a smaller value $-b_2$. When braced, both these bows have draw distance b and so have required a force $k(b + b_{1,2})$ to brace. **d** The drawn bows of draw length x_0 require draw weight $F_{1,2} = k(x_0 + b_{1,2})$. **e** Graph of force F versus draw distance x showing how this simple model can explain why the recurve bow has greater initial draw force and greater draw weight than the self-bow

Now we can plot a graph of the forces that are required to brace and draw the two bows—see Fig. 7e. Note that the initial force needed—when the archer begins to pull on the braced bow—is greater for the recurved bow than for the nearly straight bow (I think of this nearly straight bow as a self-bow). This difference in initial draw is observed for real bows. The draw weight is F_1 for the recurve bow and F_2 for the straight bow, so a recurve bow has a greater draw weight than a straight bow of the same length (assuming the same force constant k). Thus my too-simple assumption demonstrates two real features of bow draw.

The assumption is not very far away from the truth for some bows, at least for draws in the region of Fig. 7e where the force lines are solid, not dashed. Other real bows get harder to draw the further back the bowstring is pulled (which is known as *stacking* in archery jargon). For modern compound bows it works the other way—the draw gets easier as the archer pulls more (this is known as *let-off*).

More juice can be squeezed out of my simple assumption, but it takes a little physics. In TA-4 you can see how an arrow's launch speed $v_{1,2}$ for the two bows of Fig. 7a and b can be calculated from it. The result is

$$v_{1,2} = \sqrt{\frac{x_0 F_{1,2}}{m}\left(\frac{x_0 + 2b_{1,2}}{x_0 + b_{1,2}}\right)}, \quad \frac{v_1}{v_2} = \sqrt{\frac{x_0 + 2b_1}{x_0 + 2b_2}}. \tag{1}$$

The first equation shows that arrow launch speed increases with both draw weight and draw length. It is tempting but wrong to consider just the draw weight, when drawing a bow, and so to think that a bigger draw weight necessarily implies a greater arrow speed, but it is not so. For example, in the next chapter we will see that crossbows have much greater draw weights than bows, but the launch speed of their bolt projectiles is only slightly greater—the reason is that crossbows have much shorter draw lengths.

The second equation shows the ratio of two launch speeds v_1 for the recurved bow of Fig. 7a and v_2 for the self-bow of Fig. 7b. We see that a larger recurve (b_1 greater than b_2), but with length and force constant k the same, leads to a greater launch speed. We have already noted the observation that short composite bows are more powerful than short self-bows; here is a derivation that shows why, in detail. Real bows have more complicated force-draw curves than the simple linear model I suggest here, but will nevertheless exhibit the same qualitative behavior.

3.2 Efficiency

You may well ask why I am just beginning my analysis of weapon efficiency here—why there is no discussion about efficiency in the earlier chapters. The answer is that it would hardly be instructive. Efficiency in the context of our ancient projectile weapons means this: it is the fraction of energy that the thrower has imparted to the weapon, which is then transferred to the projectile. In other words it is the ratio of [mechanical energy taken away by the projectile] to [mechanical energy transferred from the thrower].[15] It would not be instructive to speak of the efficiency of an atlatl, or sling, or boomerang because the mechanical energy that is generated by the thrower is immediately transferred through the weapon to the projectile. Here we are interested in determining the energy that the thrower stores in the weapon and which the weapon transfers to the projectile. Thus, a bowstring is drawn (the archer transfers potential energy to the bow) and released. The bow then transfers the stored potential energy to the arrow. The ratio of those two energies—kinetic energy of the arrow divided by initial potential energy held in the drawn bow—is the bow's efficiency.

There are similar calculations to be done in Chap. 5 for crossbows, and in Chaps. 6 and 7 where we investigate various forms of mechanical siege engine,

[15] Please note that I am not here referring to the energy expended by the thrower during a throw. That energy includes the energy necessary to move the thrower's limbs, contract their muscles, breathe, etc.—it is the metabolic energy of the throwers actions, which can easily be several times larger than the mechanical energy, but it is much harder to estimate and, more importantly here, is not a feature of the weapon.

but no such calculation can sensibly be made for our earlier, simpler weapons. The energy of a thrower's arm moves the atlatl and the dart that she holds. The atlatl does not store the thrower's energy, but rather transfers it almost immediately. I suppose we could argue that a little mechanical energy is lost if the atlatl flexes, but that isn't the same thing.

My calculation requires a more detailed physical model of a bow than the simple $F = kx$ idea, because we need to take into account the mass of each limb of the bow, as well as the arrow mass. The bow is now considered to consist of a stiff riser and two stiff limbs of the same length, as shown in Fig. 8a and b. (Readers not interested in the technical derivations of TA-4 can ignore most of the labels.) Riser and limbs are connected by springs which stretch when the bow is drawn. This model turns out to be a good approximation of a real bow, a very good approximation of a crossbow, and an excellent approximation of a ballista.[16]

Bowstrings

Historically bowstrings have been made from almost anything pliable that has high tensile strength. Commonly plaited fibers were used: hemp, flax, nettles, even bamboo. Sometimes the fibers were soaked or chewed prior to braiding or plaiting, to improve pliability. There are different styles of braiding; a common one today and in medieval Europe is the *Flemish twist*. The Japanese longbow string (*tsuru*) was made from hemp, bound with *kusune*, a type of resin. Turkish and Arab bowstrings were often constructed from silk, or mohair. Rawhide was sometimes fashioned into bowstring. Another bowstring material: animal sinew. The favored part of the animal was the back or leg. Intestines also had the requisite strength (think of *catgut* violin strings, actually made from sheep intestines). The oldest extant bowstring belonged to the 5,300-year-old Ötzi, the Neolithic mummy, found in 1991 high in the Alps (in the Similaun Glacier—hence his name "the iceman" and his good state of preservation). His bowstring was made of sinew of some as-yet-unknown type.

It is possible that Stringfellow is a medieval occupational surname referring to bowstring makers [22]. A couple of the many bowstring videos available online are included in the Videography at the end of this book.

[16] Stiff limbs meaning, for the idealized bow of Fig. 8a, that the riser and limbs do not flex *at all*—physicists often make this assumption, because in many cases it is a good approximation to reality and it greatly simplifies calculations.

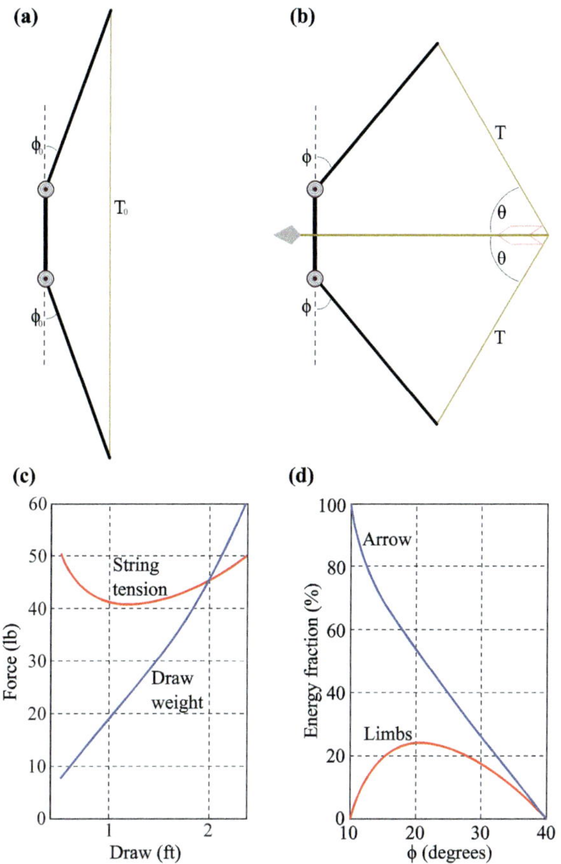

Fig. 8 Spring bow. **a** The limbs and riser are assumed to be stiff, and are joined by two identical springs. Here we see the braced bow. **b** The bowstring is drawn back to the draw length; both limb angle ϕ and string tension T have changed. **c** Bowstring tension and draw force versus draw distance for a bow with 60 lb draw weight. **d** Arrow and limb energy fraction versus limb deflection angle ϕ, for an idealized, massless bowstring

For such a bow we can set up and solve its equations of motion and the results are shown in Fig. 8c and d. Bowstring tension varies with the draw, being highest in the brace position and draw position, and less in between when the string is being drawn back, or when the arrow is being released. Draw weight now is no longer quite linear, but is almost so. The main result of the calculations is plotted in Fig. 8d: the energy fraction of different components is plotted as a function of the limb deflection angle ϕ. For this "spring bow" the full draw angle ϕ_0 is 40°. When the bowstring is released then the deflection angle returns quickly to the braced value (here $\phi_b = 10°$). Note that the fraction of energy contained in the limbs increases at first as they accelerate toward the brace position, but then fall to zero. The arrow energy increases until, when released, it carries away all of the energy that was initially stored in the drawn bow.

This idealized spring bow is perfectly efficient. It turns out that perfect efficiency is more general, and applies for many different shapes of bow [19], if we assume that the bowstring doesn't stretch (is inelastic) and is massless (weighs next to nothing compared with arrow weight). The first of these assumptions is pretty reasonable—bowstrings do stretch but not a lot. The second assumption is not good for the bows that we are interested in—those of classical antiquity and the Middle Ages. Their hemp or sinew bowstrings needed to be quite thick to withstand the tension caused by the draw weight, and so had a significant weight compared with that of the arrow and bow limbs. Bowstring weight reduces bow efficiency from 100%. Why?

Consider what happens at the instant that the arrow is released—to fly skyward and fall into a French knight at Agincourt or a Mongol at Ain Jalut. At this instant the bowstring is straight, in the braced position. A massless bowstring would stop there because it has no inertia, and so the bow is perfectly efficient. But a string with mass has inertia (that is what we mean by mass) and so it cannot stop instantaneously—it moves past the braced position toward the grip. In this position the string tension is pulling it back toward the brace position, so the bowstring and limbs end up vibrating, as the string moves back and forth, homing in on the brace position and eventually stopping there. This vibration diverts some of the initial bow energy from the arrow, thus reducing efficiency.

Even so, it is reckoned that ancient bows were nevertheless quite efficient machines—perhaps 70%. (Modern dacron bowstrings are very thin and light, and so the bows of today are very efficient—90% or so.) There is reason to suppose that strongly recurved bows are more efficient than self-bows or other more-or-less straight bows, for reasons that become apparent in Fig. 9. Two bows with the same string length are fired. When the bowstrings return to the brace position—at the instant the arrows are released—the strings vibrate, due to their inertia. Most of the straight bowstring vibrates, but less of the recurve bowstring, because the ends of the recurve bowstring are held against the bow. Smaller vibrating mass means more efficiency. Voilà.[17]

[17] See also Nieminen [23] for a (different) analysis of bow efficiency.

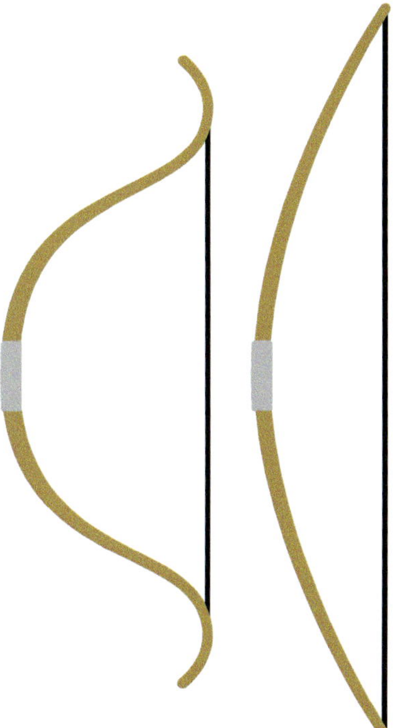

Fig. 9 Recurved bowstring length at brace height is effectively less than that of a straight bow, with the same actual bowstring length

4 Longbows

There are two longbow versions which stand out historically—those of the English and the Japanese.[18] The English longbow is famous in the Western world for what it did on the battlefields of the Hundred Years' War but not, so it seems to me, for what it was. By contrast the singular *yumi* longbows of Japan should be better known in the West precisely for what they were. Both impress—historically and technically—and yet couldn't really be more different.

4.1 The English Longbow

It is often said that the English were first introduced to the longbow during the wars of the late thirteenth century, when King Edward I ("Longshanks") sent

[18] There were other longbows, much less well known, such as the Nubian, French, and Spanish.

his armies into Wales to conquer the mountainous strongholds of Llywelyn ap Gruffudd, the last independent Welsh prince. It seems that there are not much data to support this view—that English longbows originated in Wales—and experts are divided. Wherever the origin, the bow was recognized as being a game-changer, and within a half-century was adopted widely and became the essential armament of the English in their subsequent long war with France.

In brief Edward III (Longshanks' grandson) claimed the French throne and invaded. Much of the fifteenth century and part of the sixteenth were taken up by the conflict; the French eventually won. The Hundred Years' war lingers in popular memory for only two reasons: Joan of Arc and the destruction of the French heavy cavalry in three major battles. We will be concerned with the battles.

England lost the war but won the major engagements, fought in northern France at Crécy (1346), Poitiers (1356) and Agincourt (1415). All three battles were brief and consisted mainly of French knights charging English longbowmen who were on foot, serried behind angled wooden stakes. The French knights were of the aristocracy, and both they and their horses were armored—plate armor, for the noblemen. The English archers were of peasant stock, lightly armored, and heavily outnumbered.

Heavy cavalry had over centuries come to dominate the battlefields of Europe, and yet in these three battles were relegated to history—this is the reason why the English longbow is famous.

But why this bow? Other bows of the time were more powerful, more technically sophisticated. (The Mongol composite bow was contemporaneous and with a longer maximum range—350 yards compared with the longbow's 250 yards [24].) First, they were powerful bows that fired arrows with heavy warheads (5-inch iron warheads for the armor-piercing bodkins). We have seen how bow power scales geometrically, so we can expect that the 6-foot English longbow would have been powerful. Medieval English archer skeletons show distinct distortions, [25] with thickened left arm bones indicating a large draw weight.[19] There has been much discussion about the draw weight of this bow, [26] which is reflected in the spread of estimated values shown in Fig. 2. The size of the bow meant that it fired a long arrow; the arrow length plus the large warhead indicates a heavy missile. This is why the arrow could penetrate plate armor at (maybe) 200 yards. (The Mongol bow could fire further but would not have been able to penetrate armor at this range.)

[19] A right-handed archer holds the bow with his left hand and draws the bowstring with his right. The left arm is subjected to more strain due to the way the bow is held. Edward II's mandate of 1363 obliged all able-bodied men to practice archery on Sundays and holidays, to build up strength, skill, and thousands of bowmen who could be called upon in wartime.

The last paragraph summarizes the basic reasons for why an English longbow arrow could penetrate plate armor at distance, but does not fully explain the weapon's success and fame. Other bows were more powerful (the crossbows used by Genoese mercenaries fighting for the French at Agincourt, for example). It seems to me that the main reason for the success of the English longbow was logistical, not technical.

It has been estimated that between 125,000 and 550,000 arrows were fired by the 5000 or 6000 English archers at Agincourt [27, 28]. The numbers were of the same order of magnitude at Crécy and Poitiers. Let's settle on 300,000 for the purposes of calculation. If each bodkin weighed between three and four ounces (a commonly quoted value for a heavy war arrow) then 30 tons of arrows fell upon the French cavalry at Agincourt. The front was a narrow one: 1000 yards between the forests at Tramecourt and Agincourt. The cavalry will have taken 15–20 seconds to close with their enemy from maximum bow range (hindered by muddy conditions). During which time the English bowmen will have been firing at their fastest rate (estimated to be 10 or 12 times per minute—a lot, but then they were well trained.) We now have the numbers to do the math: say 5,000 archers firing once every 5 seconds for 20 seconds means that the French knights were looking at 20 arrows *per yard of front.* I don't know the French word for "Yikes," but I can imagine it was uttered a few times on that day.

We know that there were two cavalry charges at Agincourt, and that most of the French dead were nobility (unusual, and a very striking feature—pardon the pun). Presumably more horses were hit than men; presumably those attacking on foot were exposed to a withering hail of arrows for a lot longer than 20 seconds.

The success of the English longbow against French chivalry during the Hundred Year's War was duly noted by military leaders across Europe and ended, for a century and a half, the hitherto dominant tactic of massed charges of heavy cavalry.

Logistics won these battles: archers were well trained in the use of the bow and in deploying *en masse.* Millions of arrows and thousands of bows were provided for them, carted across northern France and delivered in good condition. The English longbow was a good weapon, but not the best bow available at the time: it was cheap and plentiful and won these battles and changed warfare—but it seems to me that the English quartermasters can take a lot of the credit.

4.2 Japanese Longbow

The yumi[20] is an outlier: it is an asymmetric recurved composite longbow. In Fig. 10 you can see just how asymmetric it is: the handle is about one-third of the distance from the low end.

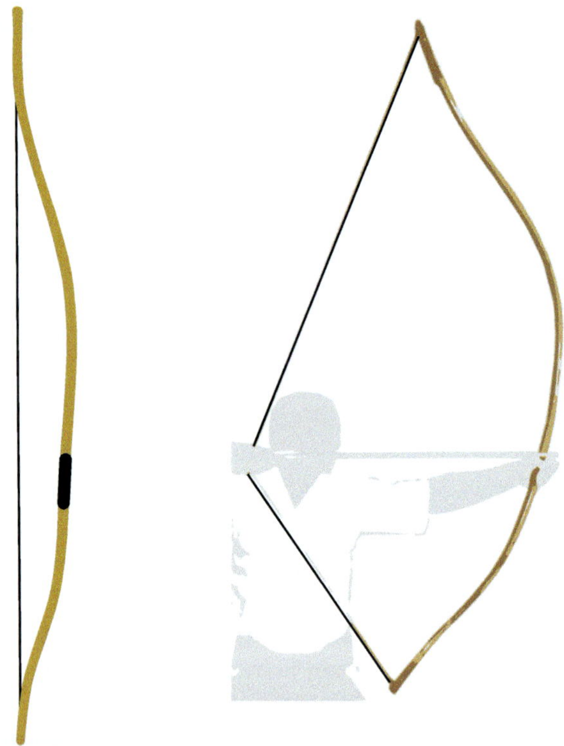

Fig. 10 The yumi Japanese asymmetric composite longbow, braced and drawn. Adapted from a Public Domain image by Bicéphal

At 7–8 feet long, the yumi is the biggest mass-produced historical longbow. It can be dated back to the third century and has evolved continuously ever since. Before the yumi came the *maruki* self-bow, lacquered to protect against the Japanese climate but otherwise an unremarkable early bow. After the yumi came firearms, as elsewhere in the world. In between lay one and a half millennia of uninterrupted use. The composite structure consists of a bamboo belly and back, and with a three, four, or five piece laminate core, and wood sides. There

[20] "Yumi" means "bow," and is the way that period sources refer to this longbow. Today it is called "wakyū," meaning "Japanese bow," because now there are many other bow varieties available to Japanese archers.

was no notch or arrow shelf in a medieval yumi. Given an ancient bow, a modern expert could tell you quite precisely the historical period it dates from ("FOUR-piece laminate core? It is Yohonhigo (Higoyumi), say 1600 CE.")

It is natural to assume that the asymmetry grew out of the restrictive movements of horse-archers, as elsewhere in the world, but there is evidence that the asymmetry predates horse-archery in Japan. A symmetric bow may have been unwieldy for a medieval foot archer[21] Another reason has to do with the physics. We have seen that, due to bowstring inertia, a bow vibrates when the arrow is released; this vibration is uncomfortable and can compromise aim. The yumi longbow has minimum vibration at the grip (i.e., the grip is located at a node, where vibration amplitude is zero).

The size and shape of the yumi meant it required an unusual draw technique, at least for mounted archers. The *uchiokoshi* procedure began with both bow arm and string arm held above the head; as the arms were lowered the bow was drawn. At full draw length the arrow nock was behind the archer's ear. For archers on foot the draw was more conventional. Unlike most archers in Asia, the thumb ring was not used in Japan.

We have seen that most composite bows were constructed using horn on the belly, but historically there were no suitable sources of horn in Japan, so bowyers used a much more common material. The bamboo was steam-bent into shape. Hardwoods like maple or beech were employed; fish glue held the components together. Lacquer protected the glue joints. Then the bow was bound with rattan or birch bark or silk. (As ever, there are excellent YouTube videos available showing modern bowyers using traditional materials to make their bows.)

The yumi has become a feature of religion and folklore, reflecting its high status as the projectile weapon of choice for samurai warriors (Fig. 11)—archery formed a significant aspect of samurai martial prowess. Yet the weapon was not exclusive to samurai; at different periods and to different degrees the *ashigaru* (peasant foot-soldiers) were armed with yumi. Certainly archers armed with yumi were a regular feature of battles over many centuries, sometimes used in serried ranks. There is a, perhaps suspect but commonly quoted, figure that indicates the significance of bows in warfare: 82% of wounds in fourteenth-century Japanese battles were caused by *ya* (arrows). (Suspect in part because many of these battles were sieges, which would produce more archery casualties than close-quarter battles.)

The ya could be quite heavy: some were 250 g (say half a pound) which exceeds that of the English longbow arrow. Of course the ya was long (33–43

[21] The average height of men in medieval Japan was 5 feet 2 inches (157 cm).

Fig. 11 Samurai with longbow, 1863. Photograph by Felice Beato. Digital image courtesy of Getty's Open Content Program

inches), which explains part of the weight. The yumi size gave it power: you can see from Fig. 2 that the draw weight was comparable to that of an English longbow. One of the arrowhead types was the bodkin, for piercing armor. So how did the yumi perform against armor?

"*The iron bolts shot from the longbows of the Minamoto archers are said to have gone crashing through the planking of the Taira junks, scuttling them as effectually as the more modern rifle-ball. As the riddled hulls sank, they left the brave warriors, swimming in a bloody sea, easy targets for the showers of relentless arrows of the Minamoto*" [29]. This is hyperbole; in reality a samurai warrior would have to be within about 30 yards of a yumi for its ya to penetrate his armor. (The maximum effective range of a yumi was quite short—90 yards or so—due presumably to the heavy ya. Maximum range is three or four times this distance.) Japanese armor varied over time, and at any time there were different grades. It is tempting to suggest that the armor of a samurai evolved in lockstep with the yumi—that they co-evolved. If so, this would be an early example of an arms race, of which more in the next chapter.[22]

In terms of sophistication the yumi is way ahead of the English longbow. Both had a heavy draw weight and both were influential in warfare. Today, both resonate with the descendants of medieval archers who used them (see Chap. 8). The English longbow was a game-changer whereas the yumi was not. In my opinion this is because the English longbow was used abroad and probably had more success in penetrating armor at mid-range[23] The French and other European nations immediately understood the effectiveness of English longbows. They got the point, as it were.

In contrast, the yumi was used largely in Japan and was not copied abroad—Japan's neighbors already possessed very effective composite short bows. It seems that there was no single "Agincourt moment"—a battle in which the yumi proved a decisive factor. This may have been due to the arms race alluded to in the last paragraph; such a co-evolution suggests that the yumi was never so very far in advance of its target armor as to be decisive.[24]

5 Archer's Paradox

An infamous (within archery circles) unknown (outside same) phenomenon, the archer's paradox expresses an interesting physical aspect of arrow release, illustrated in Fig. 12. This figure shows a cutaway view of a bow, in the plane of the arrow as it is being released. The D shape is the bow riser or grip. When fully drawn the arrow points to the target, but does not remain so when the bowstring is released, due to the finite width of the riser, as you can see. Yet real arrows do not fly off in the wrong direction as this figure would suggest—this is the paradox.

To reconcile the paradox, we must recognize that there is something wrong with the model of arrow release implied by Fig. 12. What is wrong? The figure does not take into account arrow spine, or stiffness. Real arrows are not perfectly stiff as Fig. 12 implies. They buckle under the force of the draw weight pushing

[22] At least one expert on Japanese armor considers this suggestion to be reasonable, at least during the early medieval period: "*The early type of armor worn by the mounted Samurai in the medieval period, known as Ōyoroi, was designed against bows, with thick lamellae and a buffer zone between the armor and the wearer in case of penetration. The two square shoulder guards you see are also designed to work as shields and catch arrows while allowing the warrior to use the bow*"—Luca Bottazzi, personal communication.

[23] Whether this success was due to better arrows or less effective armor, I leave you to decide. Again, please let the author know if you have an informed opinion on the matter.

[24] I am grateful to Luca Bottazzi for diligently answering my many questions about Japanese medieval archery in general and yumi in particular. Interested readers should consult his *Gunsen History* website. See also [30, 31].

Fig. 12 The archer's paradox. The bowstring lies in the plane ***ab*** (perpendicular to the page). The target is along the line ***ac***, which is the arrow direction at full draw. But at release, when the rigid arrow parts from the bowstring, its line is a′c′. So how does it fly straight to the target?

the arrow forward. In fact they buckle outward, such that the arrowhead stays on the line ***ac*** (and not ***a'c'*** as the figure shows). In other words, the arrow shaft bends so that the arrow is on the line ***a'c***. The reason why it buckles outward (instead of, say, up or down) is because the arrowhead is heavy and its inertia makes it resistant to being thrown off course—it stays on the line ***ac***.

But, you ask, what happens when the fletchings reach the riser? Why don't they collide in such a way as to throw the arrow offline? Again, the resolution is arrow spine. The shaft is elastic, so that after it deforms due to the buckling force of the bowstring, it straightens up again when that force is removed. In fact, it flexes the other way, vibrating back and forth during flight. This waggling flight—and the bending of the arrow as it slides past the rise—is beautifully captured in several recent slo-mo videos, which show how nature resolves the Archer's Paradox quicker than reading this section.[25]

The arrow waggles in flight at a specific frequency that depends on the shaft material and shaft thickness. For the fletching to slide past the riser as easily as possible, we want the arrow to have flexed half a cycle at that instant. That is, when the fletching is adjacent to the riser we want the arrow shaft to have recovered from the buckling force and flexed in the opposite direction. But the time it takes for the arrow to move from draw position (arrowhead near the riser) to release position (fletching near the riser) is dependent on the bow characteristics (bow shape and draw weight). This implies that an arrow must be matched to the bow that fires it, just as bullets must be matched to a gun.

An arrow shelf cut into the riser will reduce the amount of flex needed for an arrow to slide past. In fact many modern compound bows are *center-shot*, meaning that there is a kink in the riser so that the arrow doesn't need to slide past. The arrow will still buckle and flex back and forth, due to the strong but brief force applied to it by the bowstring upon release, but it does not have to flex at a frequency that depends on bow characteristics—it does not need to match the bow in the way that is necessary for traditional bows and arrows.

---

[25] See, for example, the YouTube video *The Archer's Paradox in SLOW MOTION—Smarter Every Day 136*.

6 Summary

Self-bows were an invention of prehistory—who knows where; probably several places independently developed this, the most important of ancient projectile weapons.

Composite bows developed early (possibly with the two-wood bow as an intermediate step)—before the Iron Age. Their construction was more complex and expensive, but resulted in a superior bow—it could be made smaller than the self-bow but with the same range and power. This feature was useful for horse-archers, who formed the backbone of the fast and mobile armies that emerged out of the Eurasian Steppes to change history. Asymmetric bows were developed for mounted archers, for ease of use.

Bow power is a combination of draw weight and draw length. The composite bow gained power over the self-bow by increasing draw weight whereas the longbow gained power by increasing draw length. Bows are efficient machines (typically 75% of energy input by the archer is transferred to the arrow).

References

1. Adrienne Mayor Quotes. BrainyQuote.com, BrainyMedia Inc, (2025), https://www.brainyquote.com/quotes/adrienne_mayor_775670. Accessed 28 Feb 2025
2. R. Insulander, The two-wood bow. Acta Boreal. **19**, 49–73 (2002). https://doi.org/10.1080/08003830215543
3. M. Kremer, Population growth and technological change: one million B.C. to 1990. QJE **108**, 681–716 (1993)
4. A. Bluet, F. Osiurak, E. Reynaud, Innovation rate and population structure moderate the effect of population size on cumulative technological culture. Humanit. Soc. Sci. Commun. **11**, 649 (2024). https://doi.org/10.1057/s41599-024-03157-4
5. M. Maschner, O.K. Mason, The bow and arrow in northern North America. Evol. Anthropol. **22**, 133–138 (2013). https://doi.org/10.1002/evan.21357
6. Encyclopedia Britannica Entry Longbow, https://www.britannica.com/technology/longbow. Accessed 09 Mar 2025
7. P. Librado et al., The origins and spread of domestic horses from the Western Eurasian steppes. Nature **598**, 634–640 (2021). https://doi.org/10.1038/s41586-021-04018-9
8. C. Ameen et al., In search of the 'great horse': a zooarchaeological assessment of horses from England (AD 300–1650). Int. J. Osteoarchaeol. **31**, 1247–1257 (2021)
9. P. Heather, *The Fall of the Roman Empire: A New History* (Oxford University Press, Oxford, England, 2007)
10. V. Bileta, Ancient Rome and ancient China: did they ignore each other? TheCollector.com (2022), https://www.thecollector.com/rome-and-china/

11. G.A. Hansard, *The Book of Archery* (Henry G. Bohn, London, 1841)
12. I. Frazier, *Invaders* (New Yorker Magazine, 2005)
13. T. May, B. Kiernan, Mongol genocides of the thirteenth century, in *The Cambridge World History of Genocide*, ed. by B. Kiernan, T.M. Lemos, T.S. Taylor (Cambridge University Press, Cambridge, 2023), pp. 498–522
14. J.J. Saunders, *The History of the Mongol Conquests* (University of Pennsylvania Press, Philadelphia, 1972)
15. S. Rumschlag, One bow (or stirrup) is not equal to another: a comparative look at Hun and Mongol military technologies. Silk Road **16**, 78–90 (2018)
16. R.A. Gabriel, *The Great Armies of Antiquity* (Praeger Publishers, Westport, CT, 2004), p.343
17. J.W. Beever, Z. Pavlović, The modern reproduction of a Mongol era bow based on historical facts and ancient technology research. Exarc (2017). https://exarc.net/ark:/88735/10291
18. C.D. Harris, O. Lattimore, A.J.K. Sanders, Mongolia (2025), https://www.britannica.com/place/Mongolia. Accessed 19 Mar 2025
19. M. Denny, Bow and catapult internal dynamics. Eur. J. Phys. **24**, 367–378 (2003)
20. B.W. Kooi, C.A. Bergman, An approach to the study of ancient archery using mathematical modelling. Antiquity **71**, 124–134 (1997). Kooi has published widely on bow physics
21. M. French, Part 5: the mechanics of simple bows. Exp. Tech. 66–68 (2006)
22. C.W.E. Bardsley, *A Dictionary of English and Welsh Surnames, With Special American Instances* (H. Frowde, London, 1901)
23. T.A. Nieminen, The Asian war bow, in *19th Australian Institute of Physics Congress, ACOFTAOS*, ed. by E. Barbiero, P. Hannaford, D. Moss (2010)
24. M. Rossabi, *The Mongols and Global History* (W.W. Norton and Company, New York, 2010)
25. A. Stirland, *Raising the Dead: The Skeleton Crew of Henry Viii's Great Ship, the Mary Rose* (John Wiley, NY, 2000)
26. M. Strickland, R. Hardy, *The Great Warbow: From Hastings to the Mary Rose* (Sutton Publishing, Stroud UK, 2005)
27. J. Barker, *Agincourt: the King, the Campaign, the Battle* [US title: *Agincourt: Henry V and the Battle that Made England*] (revised and updated ed.) (Abacus, London, 2015). ISBN 978-0-349-11918-2
28. C.J. Rogers, The Battle of Agincourt, in *The Hundred Years War (Part II): Different Vistas*, ed. by L.J.A. Villalon, D.J. Kagay (Leiden, Brill, 2008). ISBN 978-90-04-16821-3
29. R.G. Dening, *The Japanese Long-Bow*, vol. 21, no. 2 (Outing Magazine, 1892), pp. 83–91
30. K.F. Friday, *Samurai, Warfare and the State in Early Medieval Japan* (Routledge, Oxford UK, 2003). ISBN 978-0415329637
31. M. Loades, *War Bows: Longbow, Crossbow, Composite Bow and Japanese Yumi* (Osprey, Oxford UK, 2019)

5

Crossbows: Nuts and Bolts

Chapter Summary The origins and development of crossbows in ancient China and medieval Europe are introduced and contrasted. Trigger mechanisms differed and are discussed. Draw weights were high compared with bows, and influenced crossbow design. Crossbow physics is mentioned, though not in detail due to its similarity with bow physics. Crossbow bolt accuracy is discussed and bolt penetration of armor is investigated. There were many ways to span or draw a crossbow string, and these are outlined. Chinese multi-prod crossbow variants are discussed and their efficiencies calculated.

"Energy may be likened to the bending of a crossbow; decision, to the releasing of a trigger."—(Sun Tzu, Chinese general, 544–496 BCE) [1]

1 Introducing Crossbows

Structurally it would have been more appropriate to delay our investigation into crossbows until the next chapter, because crossbows are physically very similar to the torsion-powered siege engines of classical antiquity. Or perhaps crossbows, being bows, should have been part of the last chapter. However, on balance I think that crossbows are sufficiently distinct to merit their own chapter; their historical trajectory, if not that of their projectiles, is different from that of bows and of siege engines.

As with bows, let us begin with a few technical terms. A crossbowman was called an *arbalist* in medieval times, and we will do the same. The arrow is called a *bolt* (there are other names, but let's not quarrel). The stave or arms—the bow—are, for a crossbow, the *prod* (or *lath*). There are two extra

M. Denny, *Slings and Arrows*,
https://doi.org/10.1007/978-3-032-08563-4_5

Fig. 1 A small central European crossbow, possibly south German, date 1584 (engraving, probably nineteenth century). Steel bow, hemp bowstring. Length 29 in. (73.7 cm.); Width $24\frac{9}{16}$ in. (62.4 cm.); Weight 6 lb. 10 oz. (3 kg). Image courtesy of the Metropolitan Museum of Art, NY, Open Access policy

components not present in regular bows: the *stock* is the part held by the arbalist and attached to the middle of the prod—it is equivalent to a rifle stock. (In medieval times the stock was sometimes called *tiller*, confusingly since the term means something quite different when referring to regular bows). The upper end of the stock has a linear groove cut along it, running lengthways, which houses the bolt prior to launching. Lastly the release mechanism is termed the *lock*.

Crossbows (a European example is shown in Fig. 1) were used across Eurasia until they were displaced by firearms. They were invented in the first millennium BCE in China, and perhaps independently in eastern Europe. The prods were symmetrical, and there was no "archer's paradox" to worry about. The main interest for us, from the technical perspective, is the Han Chinese crossbows of classical antiquity and the steel crossbows of medieval Europe. This weapon was used elsewhere (southern and eastern Asia, the Arab domains) but was not considered especially important—a niche munition.

2 Origins and Historical Trajectory

The earliest archaeological remains of crossbows come from Bronze Age China, and it is likely that this important branch of the bow family were first developed in China, in the Warring States Period (perhaps inspired by booby traps used for hunting animals). Cast bronze parts from the trigger/lock mechanism have been found in large numbers; the oldest date to 650 BCE. The crossbow grew in military importance, peaking during the Han Dynasty (206 BCE to

220 CE—recall that it was to the Han emperor that Marcus Aurelius sent emissaries). Between a third and a half of Han armies consisted of arbalists. This was the time and place where crossbows were at their most significant in world military history [2].

Crossbows have been found elsewhere in early archaeological and historical records. Thus the Ancient Greeks made use of an unusual crossbow that they called *gastraphetes.* This odd word may sound like the name of a trendy eatery and, in fact, there is a linguistic connection. It translates as "belly releaser" because of the manner in which the bow was spanned: the stock was placed on the ground and the arbalist leaned on the other end with their belly, cocking the crossbow by using their weight. This action is more easily understood by watching it than by describing it, so I recommend interested readers search for gastraphetes on YouTube—there is quite a lot there concerning this enigmatic weapon (See the Videography).

Enigmatic because there are very few references to it in the ancient records. It likely arrived in Greece (or was developed there independently—there is no consensus among historians) in the fifth-century BCE. From this dating of the earliest European crossbow we see that it is possible it may have come from the east, ultimately from China. Both early Chinese and gastraphete bows were composites. On the other hand the trigger mechanisms were quite different.

Roman ballistas and other early siege engines are likely descendents of gastraphetes, though the Romans did not make much use of the crossbow (or *manuballista* as they called it). Indeed the crossbow all but disappeared from European warfare during the first millennium CE (but see [3]). It was used for hunting, where it had some advantages over the bow (principally, it could be held in the cocked or drawn position for extended periods) but only became of military significance during the Crusades. Of course it evolved, [4–6] and did so separately from the Chinese crossbow. Earlier (tenth century CE) European crossbow prods were made of hardwood; by the thirteenth century they were composite and by the fourteenth and fifteenth they were made of wrought iron or mild steel, with extremely heavy draw weights (many exceeding 1,000 pounds). If the crossbow did not enter European arsenals from the east, figuratively speaking, then perhaps Europeans simply picked up their hunting crossbows for military use in the many sieges that characterized the Crusades. However it happened, crossbows made a comeback in early medieval Europe. At first they were considered unsuitable for use against Christians, due to the devastating injuries they caused [7, 8] (Pope Innocent II banned their use against Christians at the second Lateran Council in 1139.) This queasiness was soon overcome, however, and the crossbow became a high-status mainstay of European warfare thereafter, until after the introduction of firearms. Their popularity increased with the introduction of metal (wrought iron or mild steel) prods—developed to increase draw weight so as to penetrate armor [9].

> ***Crossbow versus Longbow***
> Crossbows generally had a much greater draw weight than bows, but a shorter draw length. This difference was particularly pronounced with medieval European steel crossbows. When we get to the physics we will see that this leads to somewhat longer ranges for crossbow bolts, compared with arrows, and to a comparable armor penetration capability. (It is a mistake to consider draw weight alone when gauging arrow range and hitting power—draw length also must be taken into account.) Crossbows were usually more accurate because they were easier to aim. The compact (30-inch prod, typically) European steel crossbows were more convenient than bows on ramparts or castle walls—hence the value of crossbows in siege warfare. They were easier to use, and arbalists were easier to train than were archers.
> Against these advantages were the following disadvantages. First and foremost, the rate of fire was so much slower than that of bows. Second, they were more expensive to manufacture. Third, the string could not easily be removed and restrung and so was susceptible to weakening in wet weather. Fourth, the number of bowmen per hundred yards of frontage was greater than that of arbalists, and so the unit rate of fire was greater (due to more archers, quite apart from the individual higher rate of fire). Lastly, crossbows with a very heavy draw weight tended to be less efficient than bows.

Meanwhile, back in China, use of the crossbow declined with the Han Dynasty. This decline is generally considered by modern historians to be due to a preponderance of more mobile heavy cavalry (a phenomenon that covered much of the Old World in the third and fourth centuries CE—recall how the Roman Army became more mobile during this period). Another reason more specific to China is that the decline of Han infrastructure led to a decline in quality and number of crossbow trigger parts.[1] The state-run armories needed to be run efficiently and well by skilled craftsmen in order to manufacture these components with the very precise tolerances that were needed.[2] Crossbow

[1] There is an interesting parallel here with Rome, a couple of centuries later. We will see that use of the ballista by the Roman army declined due to the complexity of its construction and declining infrastructure.

[2] I am grateful to Prof. David Graff for answering many questions about the crossbows of Ancient China—this interesting snippet about trigger production is from him.

usage partially revived in the Tang Period (seventh to tenth centuries) though as junior partner to the bow. The crossbow disappeared as a military weapon only after the introduction of gunpowder weapons, as elsewhere.

It will prove instructive to compare the steel crossbow of medieval Europe with the Han crossbow—different branches of the same tree, though separated by five thousand miles and fifteen hundred years. We begin in China, because in that large and technologically advanced (for the day) country the crossbow (*nu*) was such an important part of warfare; more crossbows were employed in Han armies than any other before or since.

3 Crossbows in Han China

Chao Cuo was a Han Dynasty writer and political figure who lived in the second century BCE, at the end of the long period when the Chinese were at war with their nomadic northern neighbors, the Xiongnu. He wrote about how best to fight them: "*The strong crossbow [jing nu] and the ... javelins have a long range; something which the bows of the Xiongnu can no way equal. The use of sharp weapons with long and short handles by disciplined companies of armoured soldiers in various combinations, including the drill of crossbow men alternately advancing [to shoot] and retiring [to load]; this is something which the Xiongnu cannot even face. The troops with crossbows ride forward ... and shoot off all their bolts in one direction; this is something which the leather armour and wooden shields of the Xiongnu cannot resist. Then the [horse-archers] dismount and fight forward on foot with sword and bill; this is something which the Xiongnu do not know how to do.*" [10].

This is why nu[3] became important weapons to the Han (see Fig. 2—they were very effective against Steppe horse-archers. Millions were manufactured, with the composite prods (typically $4\frac{1}{2}$ ft long—or wide, as they were held horizontally) made of hardwood (often mulberry), horn and sinew, probably with hemp bowstrings and with bronze trigger mechanisms that evolved over the centuries (several times—historians recognize at least three distinct types they label A, B and C, of increasing sophistication) and bronze arrowheads.

Thanks to these metal components we have a lot of archaeological remains to attest to the importance and age of crossbows (crossbow parts were found, for example, along with the famous terracotta army guarding the tomb of a third-century BCE emperor's tomb [11]). Crossbows remained an important, hi-tech

[3] "Nu" also meant "fury" and these homonyms led to wordplay among Chinese writers, especially given the furious sound of many crossbows being fired at the same time, as occurred in battle. "Xiongnu" translates as "furious slave"

Fig. 2 Chinese arbalist. Wood block illustration from *The Essentials of the Military Arts*, completed in the early Song (1043) by Zeng Gongliang, Ding Du, and others at the order of Emperor Renzong. Public domain image

and high-status weapon for fourteen hundred years after the demise of the Han Dynasty, to the extent that for most of this time it was banned from private ownership (as it is today) to prevent possible rebellion and leak of technology. There were many types of crossbow, including smaller, lighter "pistols" for cavalry and heavier machines for sieges (I will examine these variants in a later section). Historians also have many records of crossbow use and, more importantly for us, of crossbow capabilities.

Crossbows were accurate—arguably more so than bows. One reason for this is that the arbalist could place an eye immediately behind the bolt and so aim along its length. Chinese crossbows had a pistol grip with a vertical trigger, as we will see, which permitted such aiming. (Medieval European crossbows were held like rifles, because of their long horizontal grips—see Fig. 1—and this too permitted placing an eye immediately behind the bolt, albeit a little more awkwardly.[4]) They had long range (Later Tang Dynasty arbalists were

[4] The ancient Greek crossbow could not be aimed so precisely; we see from videos of reconstructed gastraphetes that they had to be fired from the hip.

supposed to hold their fire until their targets closed to 225 meters—which suggests an effective range of at least that.) Their rate of fire was slow (perhaps one or two shots per minute) compared with that of bows, because of the time it took to span the crossbow. (For European crossbows, machines were often needed due to the high draw weights, as we will see in a later section. Han crossbows could be spanned by hand or with simple devices such as a belt with hooks, which took less time.) These crossbow characteristics affected tactics in battle, and the nature of Han warfare.

Early in the Han Period, and in the interregnum preceding it, warfare had become characterized by entrenchments and field fortifications, and this fact is attributed to crossbows [12]. As a consequence battles often became bogged down in stalemate (e.g., Changping, 260 BCE) Arbalist formations needed to pursue *countermarch* tactics, well known two thousand years later in the Napoleonic Wars. Thus, it took some time to reload the muzzle-loading flintlock muskets of the period and during this interval the musketeer was vulnerable to enemy fire and was defenseless. To mitigate this vulnerability infantry would form into three lines; the front line would fire, then the third line (at the back) passed through to the front and fired, while the others reloaded. In this way fire was maintained, the loaders were somewhat screened, and the lines advanced.[5] Similar countermarch tactics were employed by Han crossbow formations, as detailed contemporary accounts testify.

In summary, the characteristics of crossbows meant that they became the favored weapon of Han China[6] in its long war with the Xiongnu, and much influenced both the tactics and strategy in this and subsequent wars [13].

4 Triggers

The triggers of crossbows from classical antiquity in China were very different from those of medieval Europe. This distinction gave rise to contrasting shapes and sizes for Chinese and European crossbows. (See Fig. 3).The Chinese trigger mechanism (*ji*) was more sophisticated, despite its arising perhaps fifteen hundred years earlier. We know a great deal about these triggers thanks to the numerous archeological remains already mentioned, and also to a Chinese/Swedish expedition of 1930 that excavated Han Dynasty structures in

[5] Countermarching required training and coordination. In Wellington's war against Napoleon in Portugal and Spain the better-trained British forces needed only two ranks when countermarching; this is the famous "thin red line."

[6] One regional military inventory from the year 13BCE (the "Donghai Military Inventory") listed 537,707 crossbows.

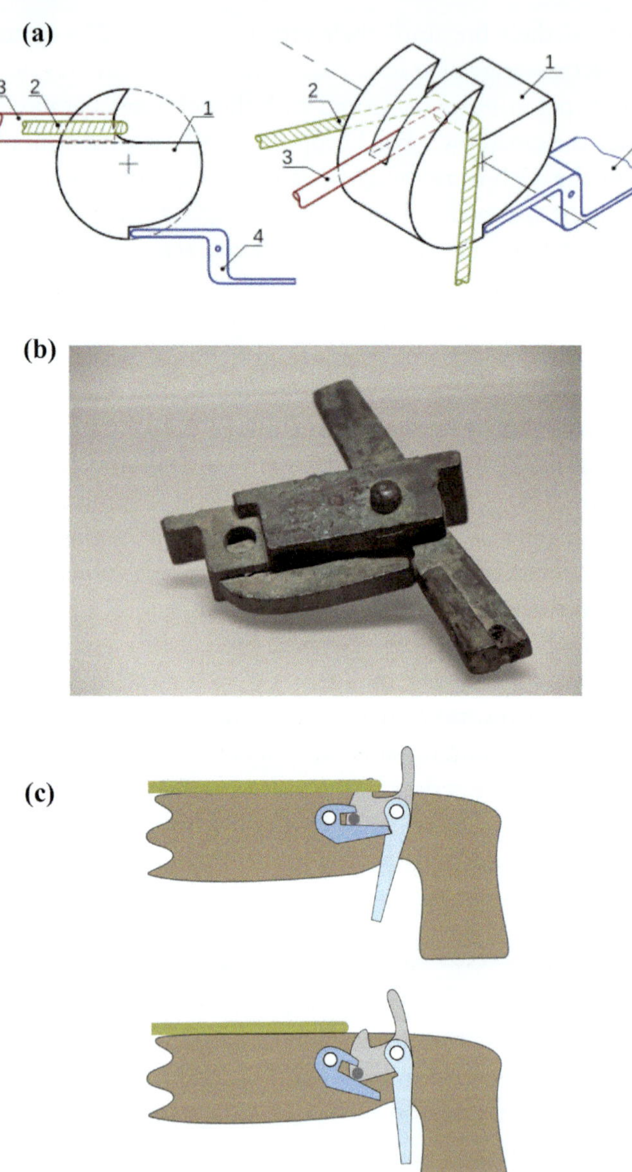

Fig. 3 European and Chinese crossbow trigger mechanisms. **a** Medieval European trigger. 1–nut, 2–bowstring, 3–bolt, 4–trigger. Image drawn by Christophe Dang Ngoc Chan. CC BY-SA 3.0. **b** Three Kingdoms Wei Bronze Crossbow Trigger, third-century CE. Photo by Gary Todd. CC0 1.0 Universal Public Domain Dedication. **c** The trigger mechanism for a spanned Chinese crossbow, before and just after release

Kansu, in northwestern China. They discovered, unexpectedly, thousands of wooden chips containing written records of all kinds of details pertaining to life in a Han Period garrison/ frontier province. These chips are now known as the *Chu Yen slips* and they are a mine of information for crossbows in general and crossbow triggers in particular.

The triggers were formed from cast bronze components. (Bronze casts much better than iron, and is stronger than copper.) They fit together as shown in Fig. 3b,c. Later in the Han Period this mechanism was protected by a bronze casing, which permitted a much greater draw weight, and made the crossbow more robust and resistant to the influences of weather and climate. (Crossbows nevertheless remained subject to some atmospheric conditions, such as temperature, because the prod characteristics changed with temperature.) The *ji* mechanism provided a mechanical advantage to the arbalist, so they did not need to exert much pull weight. This feature had two consequences. First, aim will have been improved, and second, the trigger was small and vertical with a grip at the end of the stock, like a pistol. The arbalist could easily aim and fire this weapon.

Contrast with the medieval European trigger, shown in figure Fig. 3a. The rotating nut was made from hard steel and the other components from mild steel. Hard steel was required because the nut was under great stress, due to the very high draw weight, and subjected to wear.

The mechanism provided no significant mechanical advantage and so to obtain such an advantage—to reduce pull weight (without which aiming would be compromised)—the trigger lever was long and horizontal. Typically, the trigger mechanism was placed about halfway along the stock, not at the end as for Chinese crossbows, so that the trigger could stretch along the lower side of the stock to the end, where the arbalist could grasp it. This provided a long lever and the necessary mechanical advantage. See Fig. 1 or Fig. 4. But the fact that the mechanism was in the middle of the stock meant that the stock was held like a rifle, not a pistol, and so the mechanics of operating the European crossbow were different from those of the Chinese bow. More significantly the draw length had to be shorter—only half the stock length instead of most of the length, as shown in Fig. 4—and this in turn meant that the European crossbow needed a much greater draw weight to obtain the same effective power. (We are about to see, in the next section, a physics analysis which shows why this is the case.) The steel prods of later medieval European crossbows enabled such enormous draw weights, and so this design worked. Of course, there was still the troublesome point of how to span such a bow, but we will get to that later—there were many mechanisms developed over the years.

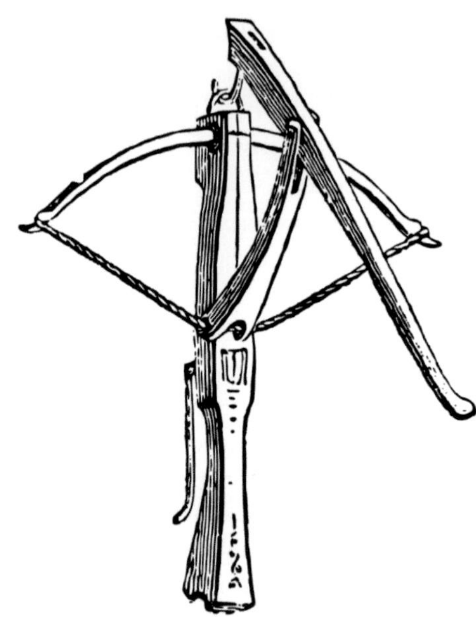

Fig. 4 Medieval European crossbow, with gaffe spanning lever. Note the trigger mechanism in the middle of the stock, and the long horizontal trigger. Author unknown, public domain image

The technical details of crossbow triggers have exercised the minds of specialists and non-specialists alike. Both triggers improved over time as the basic working designs were tinkered with and new ideas realized. I recommend that the interested reader views some of the many YouTube videos discussing, building, and analyzing the two types of mechanisms I have outlined here—see the Videography.

5 Crossbow Mechanics

Data from the Chu Yen slips tells us about Han crossbow draw weights, bolt weights, and the connection between the two [14]. It seems that the Han thought that there was an optimum weight for a bolt, for a given draw weight. The Chu Yen data suggest that, across all draw weights, the ratio should be about 1,900. That is to say, for the most common crossbow draw weight (390 lb) the best choice for the weight of the bolt it loosed would be about 0.2 lb.[7]

[7] This bolt weight of 0.2 lb is about 3.2 oz (ounces), or 1400 g (grains—a common unit today and in the past for arrow and bolt weights), or 93 g (grams) or 6 liang (that is Han liang, you understand, not modern Chinese liang). Just thought that you would like to know. (Why won't everyone switch to metric?).

Given this fixed ratio, we show in TA-5 that draw length is the single most important characteristic of crossbows, in the sense of determining performance. We saw in the last chapter that the product of draw weight and draw length determines a bow's power, and the same is true for crossbows. Draw length alone determines other parameters for Han crossbows. Consider several Han crossbows which require the same energy to span. Everything else about them is different—draw weight, prod dimensions and composition, bolt size and weight, stock length. For all of these, the bolt release speed is proportional to the square root of draw length, and is not very sensitive to other parameters. The maximum range is proportional to draw length, and is likewise not very sensitive to other parameters.

The Han preference for a constant ratio of draw weight to bolt weight tells us that they thought there was an optimum weight for bolts, for a given crossbow. Why should this be the case? Considering only internal ballistics, the best choice is lightweight bolts, because for a given energy they fly faster and therefore further than heavier projectiles. But terminal ballistics require target penetration, and this is achieved more successfully with heavier bolts. It is a projectile's momentum (mass multiplied by velocity) that is the key factor for target penetration because Newton's Second Law tells us that force is the rate of change of momentum with time. Thus a bolt that penetrates a target is decelerating and so is exerting force on the target.

You can do damage to a target without penetrating it, of course, and in such cases, energy deposition is more important than momentum. Let us suppose that you hit a bad guy with a baseball bat[8] and the bat energy sends shock waves out from the point of impact, tearing tissue and breaking bones. The greater the bat energy, the larger the volume of tissue and bone that is damaged. But this type of damage is not so important to us because we are interested in penetrating armor. Hit a bad guy in plate armor with a baseball bat, and he won't like it, but it will do less damage than hitting him with a crossbow bolt (with lower energy, but greater momentum concentrated at the smaller point of contact).

Thus there is a best choice for bolt weight for any given crossbow, if your target is armored. In TA-5 you can see a simple derivation of the following equation relating momentum p and target range R: $p^2 R = \text{constant}$. I expect this equation to be approximately true for any given crossbow (or bow) that fires projectiles with the same energy each time. It tells us, roughly, how penetration range depends upon armor strength. Here is an example to show how this works. You are an arbalist firing at an armored enemy (your baseball bat just

[8] Please do not try this at home—leave hitting bad guys with baseball bats to experts.

didn't do the job). Say a particular mail armor requires your crossbow bolt to have momentum $p = 7$ kg ms^{-1} to penetrate it from a distance $R = 100$ m. The enemy you face today wears leather armor requiring only $p' = 4$ kg ms^{-1} momentum, which means that you can hope to skewer him out to a range $R' \approx 300$ m.

6 Spanning a Crossbow

Because of their large draw weight, compared with bows, crossbows cannot be spanned using just arm muscles. For moderate draw weights the bowstring can be drawn by hand as follows. The arbalist places a foot in a stirrup that is attached to the bow end of the crossbow stock. The stirrup is on the ground, and the stock is vertical. The arbalist can now grab the bowstring with both hands and pull it upward until it engages with the trigger mechanism. Back muscles are used in addition to muscles of both arms—this method works for drawing weights up to a couple of hundred pounds.

Moving up to heavier draw weights of about 300 lb, a spanning belt can be used. Hooks are suspended from a stout leather belt around the arbalist's waist; these hooks are attached to the bowstring—the rest of the action is just as before, using a stirrup. Here, back and leg muscles are added to arm muscles. Han crossbows were spanned using a belt and stirrup, typically, with the arbalists standing or lying on their backs. (Ancient sources noted that it was not advisable for crossbow units to operate on swampy ground!)

We have seen that medieval European crossbows required higher draw weights and shorter draw lengths, and so spanning equipment was needed that provided mechanical advantage (see Fig. 5a–c). The simplest and oldest was the gaffe lever. This λ-shaped lever *pushed* the string onto the trigger, as suggested in Fig. 5d. It could span crossbows with draw weights up to about 400 lb. The more refined goatsfoot lever (Fig. 5a), introduced into European countries in the thirteenth to fifteenth centuries CE (dates here vary with source and region) *pulled* the string onto the trigger. The action (Fig. 5e) applied increasing mechanical advantage as the bowstring was stretched toward the trigger, and could span crossbows with draw weights up to about 500 lb.

Steel prods required significantly higher draw weights, and so needed altogether more sophisticated spanning devices. The windlass (Fig. 5b) was introduced around 1400 CE. To span a crossbow it was fitted over the butt end of the stock, with two hooks attached to the bowstring. The arbalist simply operated the winding gear handles and could span crossbows with draw weights up to about 1,200 lb. The increased capability came at a cost: compared with

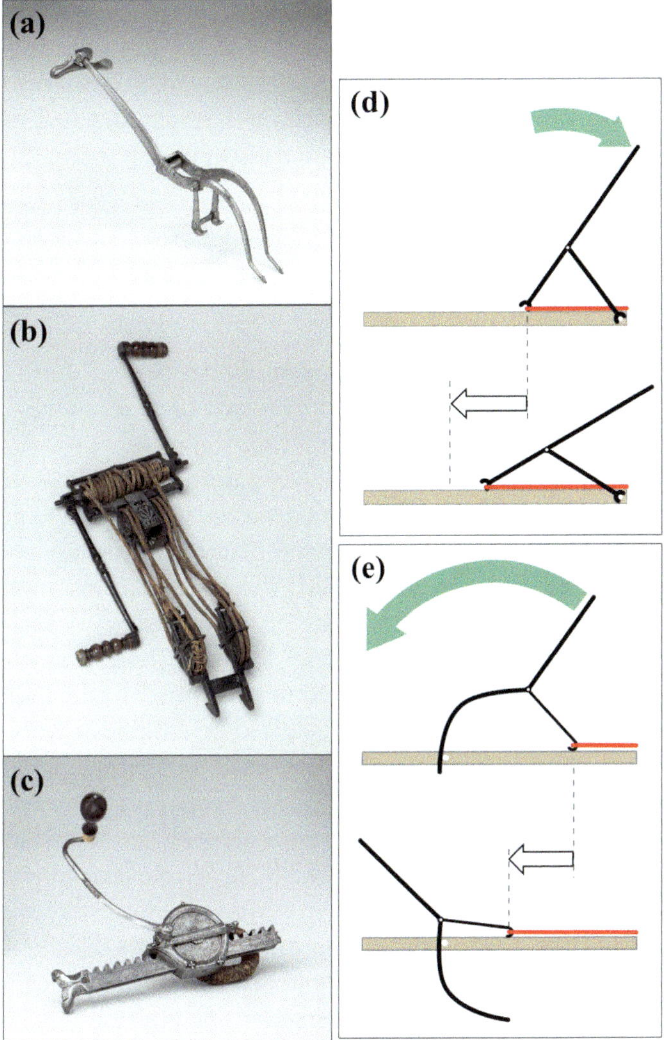

Fig. 5 European crossbow spanning tools. **a** Goatsfoot lever, probably Spanish, late fifteenth or first half of the sixteenth century. **b** Windlass, Belgian or Dutch, dated 1741. **c** Cranequin, German, ca. 1575–1650; dated 1562. Public domain photos courtesy of Metropolitan Museum of Art Open Access policy. **d** Gaffe lever operation schematic. Here the crossbow is viewed from the side, with the bow and trigger mechanism not shown, for clarity. The front end of the stock is to the right, and so the bowstring (red line) is stretched to the left. **e** Goatsfoot lever operation schematic

the spanning belt method, or simple levers, it took a long time to span a crossbow using a windlass. Same for the *cranequin* ("like a crane") shown in Fig. 5c, which was introduced in the fifteenth and sixteenth centuries CE: it was slipped over the butt end of the stock, with hooks attached to the bowstring, was operated by a winding handle and could also span 1,200 lb draw weights. But it took a long time.

7 Variants

We have seen that the use of crossbows in China peaked during the Han Dynasty of classical antiquity, when Rome was also at its peak.[9] Thereafter their use declined, but revived again under the Tang and Song. A significant Tang (seventh to tenth centuries CE) innovation was the *multi-prod* family of crossbows, a class of projectile weapon that was used mostly in sieges.

The multi-prod bow was a strange beast—or, rather, a menagerie of distinct but related creatures. They were collectively successful attempts to increase the power and range of a crossbow by arranging, in various ways, for more than one prod to combine in transferring energy to the projectile. With greater draw weights, each of these machines had to be spanned by a windlass (or the efforts of several men) and the trigger released by striking it with a wooden mallet. The projectile itself was usually larger than those of standard hand-held single-prod crossbows, and designed for a specialized purpose. For example, one variant shot a large and strong bolt into castle walls that would stay in place and assist besieging troops to scale the walls. A double bow might have two prods side-by-side, or back to back. Three-prod weapons were widely used. There were also large single-prod, or two-prod, crossbows that fired two bolts or a bucketful of bolts. All these weapons were fixed on wooden beds and must have required considerable tinkering and incremental improvement to get them to work properly. But work properly they did, given the reports of historical records and the fact that some of these designs persisted for centuries. I think that they must all have been niche products developed for very specific (siege) applications. To the modern engineer they come across as test-beds, or prototypes, or the creative product of military research corporations (I suppose the Tang equivalent would have been provincial state armories) tasked with

[9] During these centuries there were four large empires across Eurasia concurrently. The Roman Empire in the West often fought against the Parthian (Persian) Empire to its east; the Parthians' eastern frontier was with the Kushan Empire, a nomadic confederation which bordered Han China. In theory a trader from lowland Scotland could travel to Luoyang in eastern China and cross only three boundaries. The Silk Road trade route arose in this time, facilitating the movement of trade goods between Rome and Han China.

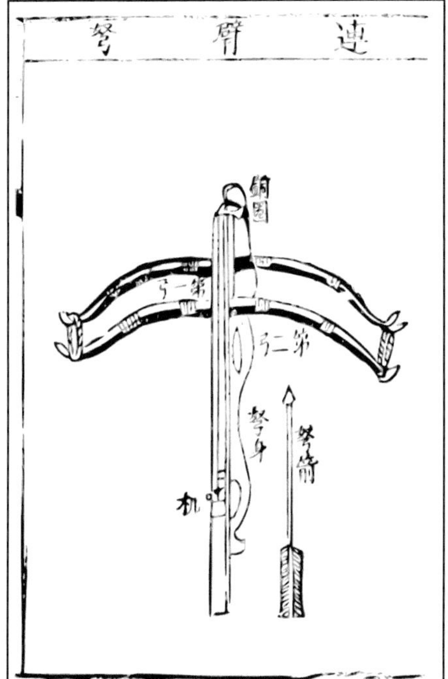

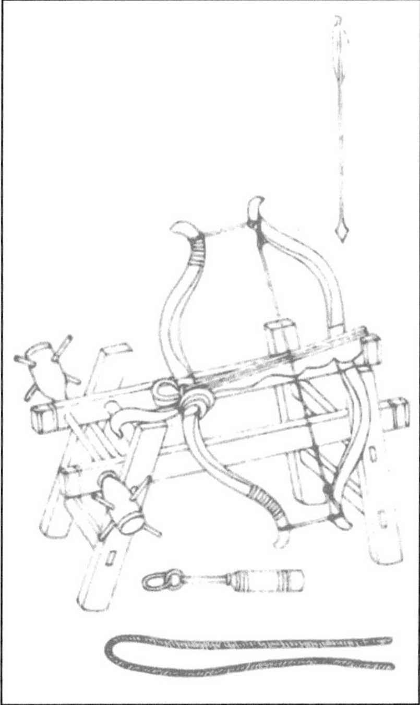

Fig. 6 A pair of 2-prod crossbows. Left: *Lian bi nu* with prods of approximately 30 inches and a draw length of half that—the size is more like that of medieval European crossbows. Right: *Xiao He Chan Nu* were useful in sieges where their extra range and power were an asset and their lack of mobility not a hindrance. The bowstring provides linkage between the prods

solving a specific problem. However they came about, they are interesting technical variants of long-standing and mature crossbow technology.[10]

Two of the simpler multi-prod crossbow variants are shown in Fig. 6. The *lian bi nu* ("linked-prod crossbow") was an unusual hand-held two-prod crossbow with the (composite) prods linked together, as the name suggests. Clearly the intent was to double-up the draw weight for the same draw length. In this case the stirrup shows that this odd bird could be spanned by a single arbalist, which suggests to me that each prod must have been of quite low draw weight.

The *Xiao He Chan Nu* ("small combined cicada crossbow") had two bows mounted back-to-back on a bed. Depending on parameters (such as bowstring length and draw weights) this bow might have increased the bolt energy by increasing draw weight or by increasing draw length, or a combination of the

[10] The author is grateful to Prof. David Graff and to 戰國春秋 of the *Great Ming Military* blog for their insights into Chinese crossbow technology.

two. A number of different cicada models exist (the name comes from their back-to-back nature, which brought to mind mating cicadas) including one with two linked bows facing forward and one facing back. It would be very difficult to analyze these bows to predict their mechanical performance in detail, and a separate analysis would be needed for each variant, so in TA-5 I have gone with a (rough) estimate of energy efficiency which serves all types, and which provides insight into the optimum number of prods. The result of this calculation is that the energy gain factor for an n-prod bow is $G_n = n\epsilon^{n-1}$, where ϵ is the energy efficiency of a single-prod crossbow. This analysis assumes that all n component prods in the n-prod structure are identical, and predicts that the energy of the bolt as it leaves the bowstring is G_n times the energy of a bolt leaving the single-prod crossbow. Plugging in realistic values for single-prod crossbow efficiencies (see the table in TA-5) we see that the best choice is $n = 2$ for moderate efficiencies (ϵ between 0.55 and 0.65). If each prod is of higher efficiency (between 0.65 and 0.75) then a $n = 3$ prod crossbow is the best choice. That these were the most common types of multi-prod bows, and the most usual bow efficiency ranges, suggest that there is some merit in the calculation. Four- or five- or higher prod bows are favored only for increasingly high (and unrealistic) efficiencies.

The energy increase for the 2-prod bow, suggested by this calculation, is about 30%; for the 3-prod bow it is as much as 75%. Treat these predictions as approximate, given the approximate nature of the calculation that generated them. There is not much historical data available concerning the actual gain in bolt energy (and so in bolt penetration or range). There are claims of over a kilometer in maximum range—well, maybe. A number of modern researchers/hobbyists/enthusiasts have carried out some "experimental archaeology" studies to estimate performance, which consist of reconstructing different multi-prod bows and seeing how they function, looking at the engineering compromises and practical considerations that are needed to make these contraptions work. Again, the interested reader can do no better than view the many YouTube videos on this intriguing subject, some of which are listed in the Videography.

There is a classic book on crossbow history [15] written by Ralph Payne-Gallwey over a century ago, that describes another Chinese innovation: the world's first semi-automatic weapon. The *chu ko nu* repeating crossbow is shown in Fig. 7, adapted from Payne-Gallwey's book. He began his account of the repeating crossbow with the words: "*Here we have surely the most curious of all the weapons I have described.*" He is not wrong.

Presumably the bolts or darts did not have flight feathers, or else they might not be fed cleanly to the stock groove. This will not have mattered because the

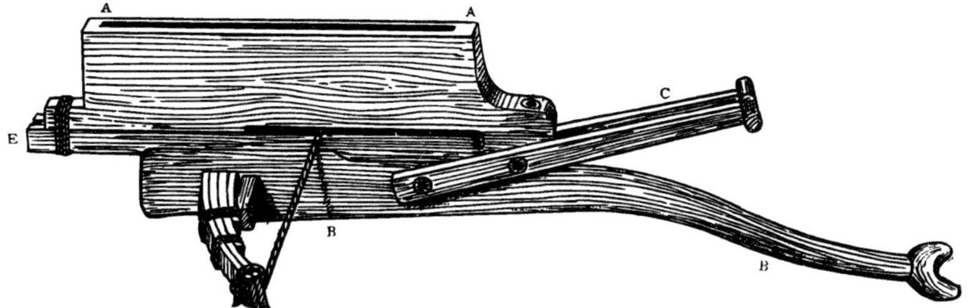

Fig. 7 A *chu ko nu* hand-held repeating crossbow. Bolts are held in the magazine (**A**) and fed into the groove on the stock (**E**). The crossbow is spanned, the bolts fed, and the trigger fired by pumping the lever (**C**) back and forth. Public domain image adapted from Ref. [15]

weapon was for close-range use only. It cannot have been very powerful given the short draw length and the fact of its being spanned by only one arm. The bolts were sometimes made more dangerous by being tipped with poison.[11] The *chu ko nu* could fire its bolts as fast as the arbalist could pump the lever back and forth (dare I call it "bolt action"?). It is of surprising antiquity: the earliest archaeological evidence dates from the fourth-century BCE, during the Warring States Period.

Due to its low power, the repeating crossbow was useful only for personal defense, or perhaps at close quarters during sieges. It might have had a range of 70 m; the fact that the basic design remained virtually unchanged for centuries (though there were variants developed) points to its efficacy [10, 16]. As always, here has been significant "experimental archaeology" and interested readers can find some example in the Videography.

8 Summary

Crossbows were an important class of weapon in ancient China, where they were invented perhaps 2,700 years ago. They were the main weapon of the Han Dynasty in their long series of wars against Xiongnu nomads. Later the crossbow fell out of favor with the Chinese as the military character of their enemies, and of warfare, changed. Crossbows revived at the end of the first millennium CE under the Tang and Song, and crossbow designs flourished into many heavier multi-prod forms to assist with siege warfare.

[11] According to the Wikipedia article *Repeating Crossbow* the poison was likely derived from the wolfsbane or monkshood plant.

Chinese crossbows spread to nearby regions, but the European crossbow of classical antiquity was likely invented independently. It was used for hunting, and formed a minor component of Greek and Roman munitions. In medieval times it became a significant feature of warfare, with the need for projectiles to penetrate the mail and plate armor of knights. The European crossbow generally had a higher draw weight and shorter draw length than the Chinese crossbow, and possessed a similar or somewhat greater ability to penetrate armor. The main drawback of crossbows was their slow rate of fire compared with bows.

References

1. Sun Tzu, *The Art of War*
2. C.M. Wilbur, The history of the crossbow, illustrated from specimens in the United States National Museum, in *Warfare in China to 1600* (1st ed.), ed. by P. Lorge (Routledge, Oxford, 2005), chapter 2. https://doi.org/10.4324/9781315234359
3. G.T. Dennis, Flies, mice, and the byzantine crossbow. BMGS **7**, 1–5 (1981). https://doi.org/10.1179/030701381790206571
4. S. Ellis-Gorman, *The Technological Development of the Bow and the Crossbow in the Later Middle Ages*, (Unpublished PhD Thesis, Trinity College Dublin, 2016). https://www.researchgate.net/publication/349867529
5. V. Foley, G. Palmer, W. Soedel, The crossbow. Sci. Am. **252** 104–11 (1985). http://www.jstor.org/stable/24967552
6. M. Loades, *The Crossbow* (Osprey, Oxford, UK, 2018)
7. M.G. Krukemeyer et al., Survived crossbow injuries. Am. J. Forensic Med. Pathol. **27**, 274–276 (2006). https://doi.org/10.1097/01.paf.0000221086.42098.72
8. C.S. Rogers et al., Crossbow injuries. J. Forensic Sci. **35**, 886–890 (1990)
9. J.F. Guilmartin, Military technology. *Encyclopedia Britannica*, 22 Oct. 2024, https://www.britannica.com/technology/military-technology. Accessed 1 Apr. 2025
10. J. Needham, *Science and Civilization in China* (Cambridge University Press, Cambridge, UK, 1994)
11. X.J. Li et al., Crossbows and imperial craft organization: the bronze triggers of China's Terracotta Army. Antiquity **88**, 126–140 (2014)
12. D.A. Graff, *Medieval Chinese Warfare*, 300–900 (Routledge, London, 2002). 0415239559
13. F.A. Kiernan, J.K. Fairbank, *Chinese Ways in Warfare* (Harvard University Press, Cambridge MA, 1974). 0674125754
14. See the online History Forum *Han Dynasty Crossbow III*. https://historum.com/t/han-dynasty-crossbow-iii.179336/

15. R. Payne-Gallwey, *The Crossbow* (Longman's Green and Co., London, 1903)
16. K.H. Hsaio, H.S. Yan, Structural synthesis of ancient Chinese Chu State repeating crossbow, in *Advances in Reconfigurable Mechanisms and Robots I* (Advances in Reconfigurable Mechanisms and Robots I, 2012), pp. 749–758

6

Vertical-Axis Siege Engines: Birth of the Machine Age

Chapter Summary The variety and importance of siege engines to ancient warfare is outlined and presented. Ballistas are described, and their origins and development are discussed in detail. A simple approximate expression for ballista projectile (either arrow or rounded stone) launch speed is derived, and predictions made for the ranges of ballistas. The "scaling rules" of ancient ballista engineers are presented and explained; scaling behavior is discussed more generally. Ballista efficiency is estimated. The growth of ballistas during the period of their domination of European siege warfare is noted. Large and small "benchmark" ballistas are analyzed and their performances presented. These predictions compare well with the claims made by ancient sources of ballista capabilities, and with the performances of modern reconstructions.

"We were perplexed and uncertain where first to offer resistance, whether to those who stood above us or to the throng mounting on scaling-ladders and already laying hold of the very battlements; so the work was divided among us and five of the lighter ballistae were moved and placed over against the tower, rapidly pouring forth wooden shafts, which sometimes pierced even two men at a time."— (Ammianus Marcellinus, Greek and Roman soldier and historian, ca. 330–400 CE *Roman History* XIX 5.6.[1])

[1] See https://penelope.uchicago.edu/Thayer/E/Roman/Texts/Ammian/19*.html.

M. Denny, *Slings and Arrows*,
https://doi.org/10.1007/978-3-032-08563-4_6

1 War Machines

Siege engines represent a significant break with the weapons that we have seen in earlier chapters because they were the first that could not be used for hunting. They were purely war machines, used for sweeping away enemy soldiers who were attempting to scale your castle walls, or for knocking down those walls, or for you to destroy their siege towers or battering rams. The ancient projectile weapons that we have considered thus far have all been personal weapons—with a crew of one, if you like. From here on we will find increasingly that crew size gets bigger as the engines get bigger. Yes, you will find small siege engines with one-man crews in this chapter and the next but, even for these, more than one person was involved in their construction. Indeed, we will see in this chapter that siege engine design and development are an early example in human history of complex machines and planned R&D programs.

The menagerie of projectile siege engines is a large one, which I will split into two families with a chapter devoted to each. My choice of how to make the division is unusual—by rotation axis—but reasonable on physics grounds. The horizontal-axis siege engines of the next chapter are stone-throwers, and they impart backspin to their projectiles. Backspin provides aerodynamic lift, which can be significant for long-range trajectories. The vertical-axis engines of this chapter—the *ballista* family—throw both arrows (or darts, bolts, javelins—I will call them all "arrows" here) and stones. They impart no spin, and so the projectile trajectories are less stable and predictable. For this reason, it seems to me, significant effort was made by artillerymen of old to make their stone projectiles as round as possible, so that they would not veer off in a random direction during flight. A more important difference is the fact that horizontal-axis engines almost always included a sling as an essential component of their structure, which vertical-axis engines could not.

A historically more natural division would have been between the classical torsion engines and the medieval gravity engines, in which case the oddball onager would have found its home alongside the ballista of this chapter, rather than the trebuchets of Chap. 7. No worries—all the major beasts in the siege engine menagerie will be discussed and analyzed either here or there.

2 Historical Trajectories

From their earliest days, the siege engines developed, improved, morphed, and evolved across the ancient and medieval world. Due to their generally rigid form, the mechanical analysis of siege engine function and performance is relatively easy (compared with, say, the sling). Combine this fact with another, that siege engines arose in civilizations that were mostly literate, and you might reasonably suppose that the history of these machines is pretty well known, with a consensus among historians. However, this is not the case. A broad spectrum of opinions exists concerning the effectiveness of siege engines, when and where they were introduced, and even their names. Siege engine physics is fairly well pinned down, as you will see, but for me the history is more elusive—a will o' the wisp rather than an illumination—and so more problematical. I will provide both physics and history, but with caveats regarding the history.

Part of the problem is terminology. Different peoples over history gave different names to the same machine, and sometimes a given people changed the name for a particular machine, or the machine changed its function while retaining the name (I am thinking here in particular of the ballista, a Greek and Roman engine that switched to stone-throwing from bolt-throwing). Compounding the problem is the fact that, in the Western world, most of the historical sources before the fourteenth century are found in religious chronicles with authors who, with a few exceptions, knew little of warfare. This fact makes it difficult to know which machine a given source is mentioning. It seems that siege engines were widespread and yet (perhaps because of this fact, because they were so well known) the descriptions are vague, as if the authors assume that their readers know much already of what they describe.

The wishy-washy nature of our historical knowledge means that even today people use different words for these machines, and so here it is necessary for me to make clear what names I am using and what machines I will be describing and analyzing. First, what do I mean by a *siege engine*? I will restrict attention in this book to projectile siege engines, omitting such devices as mobile assault towers or battering rams, and such like. Here a siege engine is a device, usually a machine with many moving parts mounted on a frame, that is used by forces besieging a walled city, or by defenders of that city. The besiegers might be trying to knock down the walls or send missiles over them to strike softer targets within. The defenders might by trying to kill their assailants with their engines, or trying to destroy their assailants' siege equipment. So far so good—no vagueness or disagreements here.

In the Western world the evolution of siege engines—the historical trajectory—took the following form: tension, torsion, traction, and then gravity. That is, the first machines fulfilling our definition of siege engines were powered by tensile forces (they were overgrown crossbows). These were then largely replaced by torsion engines that were powered by twisted skeins of horsehair or sinew fibers. The traction trebuchet siege engines were introduced from the East, and these dominated until the development of gravity-

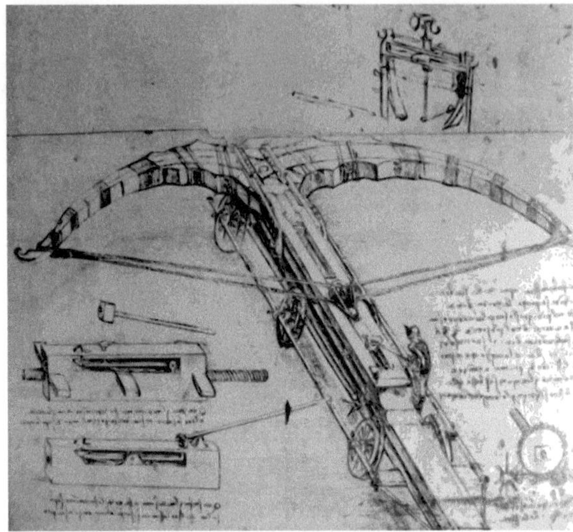

Fig. 1 Leonardo da Vinci (1452–1519) proposed a giant crossbow (never realized) for sieges. During his lifetime cannons were becoming the dominant siege weapons [1]

powered machines, which took over the heavy lifting, literally and figuratively. In China the historical trajectory was both simpler and better known (due to more knowledgeable sources): they developed traction-powered siege machines early on, skipped torsion engines entirely, and took up gravity engines introduced from the West.

Now for the engines themselves. We have seen already how crossbows were cobbled together in different ways to form larger and more powerful machines for siege purposes. Leonardo da Vinci conceived of a giant crossbow siege engine, shown in Fig. 1, though in the West tension engines reached their peak in classical antiquity, nearly fifteen hundred years before Leonardo's time. I will not describe tension engines further in this chapter. Torsion engines include the *onager*, or "wild ass" due to its kick, as we will see. This one-armed catapult is often portrayed with a spoon arm, mounted on a heavy frame. The ballista of this chapter is a two-armed machine, best known for throwing darts and for being standard equipment in the Roman Army.

The *traction trebuchet* (a modern term) is a lever powered by human muscles—lots of them—and was the main siege engine around the Old World for several centuries. It is sometimes called a *mangonel*, which is an older word and a generic term for any stone-throwing catapult. Here I will not use this word, as it sows confusion, being sometimes employed to describe our big-boy gravity engine, the *counterweight trebuchet.* Because the axis about which their projectiles rotated was horizontal, both of these machines will be described and analyzed in Chap. 7.

There were many variants and hybrids, some of which will be mentioned in passing, but they are mechanically very similar to the main subjects considered here and in the next chapter, and so do not merit more than a reference.

3 Catapult R&D, and War

The earliest representation of a catapult may or may not (scholars quibble over this) be depicted in a ninth-century BCE stone relief, from the Assyrian capital of Nimrud, in modern Iraq. Then there was an eighth-century BCE king of Judah who, according to the Old Testament, oversaw the construction of a catapult device of some kind.[2] There is a fifth-century BCE Indian text that refers to stone-throwing catapults [2]. Little or nothing is known about these ostensible machines, and so I take these statements as indicating an earlier origin to siege engines than that discussed here. I will restrict attention in this section to catapults of the Mediterranean world, about which we know much more. Our story begins in the Greek city of Syracuse (Sicily) in 399 BCE and so here I will use the word "catapult" to refer generically to projectile siege engines of classical antiquity in the Mediterranean region. The word comes from the Ancient Greek *katapeltēs* meaning to "hurl down," because catapult projectiles could penetrate both a shield (*pelte*) and the soldier holding it.

3.1 Dionysius and Phillip

Many historians consider that the catapult developed rapidly under the patronage of Dionysius I, tyrant of Syracuse, who was anticipating a long war with mighty Carthage. In 399 BCE he put together what today would be called a research and development program. He "*...gathered skilled craftsmen, commandeering them from the cities under his control and attracting them by high wages··· his purpose was to make weapons in great numbers and every kind of projectile··· the catapult was invented at this time···, since the best craftsmen had been collected from everywhere into one place. The high wages as well as the numerous prizes offered to the craftsmen who were judged to be the best stimulated their zeal. Moreover, Dionysius circulated daily among the workers··· and*

[2] "*Uzziah made in Jerusalem engines invented by cunning men, to be on the towers and upon the bulwarks, to shoot arrows and great stones withal.*" Chronicles 26:15.

rewarded the most zealous with gifts and invited them to his table".[3] (See [3] p. 771.)

The goal of Dionysius was to invent novel weapons of war, and his team of specialists were engineers from around the Greek world. We do not know their names, but we do know many of the later engineers who carried forward catapult development and who, fortunately for us, wrote about it. Dionysius' preparations included expanding his army (to over 80,000 men—no small number, he was no petty tyrant) and navy, making and importing arms and armor, and building walls. His new weapons included larger ships (*quadriremes*, with four banks of oars instead of three in the older *triremes*) and new catapults. In fact he engaged in two wars with the Carthaginians, in 397-396 BCE and 392 BCE, resulting in a win and a draw.

The earliest Greek catapults, for use in sieges, were enlarged crossbows (*gastraphetes*) placed on a frame and spanned with a winch pull-back. Dionysius' engineers moved away from tension to torsion, giving us (actually, selling us) the two-armed torsion machines that changed siege warfare. In Fig. 2 we see a typical two-arm catapult of the type introduced by these engineers; though this illustration is of a later version it serves to show three of the four characteristic features of this type of engine. First and foremost note the two vertical skeins of twisted sinew or horsehair.[4] These skeins were made of many cords and so were more powerful than tension engines, with their single cord—the bowstring. (Horsehair has a tensile strength that is similar to that of steel—overtorque the spring and the bow arm will break, not the spring material.) For the new torsion engines there was one skein spring per bow arm. Readers with engineering experience will wonder how well this arrangement could work: the skein springs were independent and were likely not quite identical, so how could the catapult operators be sure that both ends of the bowstring were being pulled with the same force (surely necessary for successful operation)? In fact the skeins could be torqued independently, to even up any difference detected by the operators, as we will see. The bows themselves were thick and rigid, so that all of the potential energy that powered the projectile came from skein torsion, not bow arm bending. (Landels discusses the spring system in some detail [4].)

Secondly we see at the top and bottom of each skein there is a washer (*modioli*, they were made of bronze) through which the cords of the skein were

---

[3] Diodorus of Sicily, Library of History, C.H. Oldfather, transl. (Harvard Univ. Press, Cambridge, MA, 1954), Sects. 14.41.3-âŁ"14.42.2.

[4] Or human hair, at a pinch. A famous and possibly even true example had the women of Carthage donate their hair for catapult skeins during the desperate final siege of their city by the Romans during the Third Punic War, 149–146 BCE. Hair was soaked in olive oil (and treated in some way with grease or fat) prior to being woven into a skein [4].

Fig. 2 Left: a ballista-type stone-throwing two-armed catapult, a lithobolos, with features described in the text that are typical of the torsion engines of classical antiquity. Public domain image drawn by Pearson Scott Foresman. Right: a more detailed view of a skein-of-cord spring, with more prominent bronze washers

threaded—see Fig. 2—then passed over an iron rod or lever and back down through the washer. Keep these washers in mind, because they will prove to be important.

Third, note the stock with a channel cut into it to accommodate a stone projectile—the earliest of these catapults fired arrows, not stones, for which the "channel" was a narrower groove, and the whole machine was smaller. There is a slider, connecting bowstring to winch mechanism, to wind up (span) the bows.

Not shown in Fig. 2 is a fourth feature commonly found in two-armed catapults: a tilt-and-swivel universal joint (*karchēsion*) connecting the stock to the frame, thus permitting the stock to be swiveled easily without lifting and turning the whole engine.[5] A universal joint thus widened the field of fire and simplified the aiming process, and the speed of aiming, which was an important factor in the operation of small, anti-personnel arrow-throwers.

Now that I have introduced to you the salient features of a generic two-armed catapult, I can convey the development and improvement of these weapons—

[5] The universal joint is generally considered to have been invented by Robert Hooke or Geronimo Cardano two millennia later.

their design evolution—over the seven or eight centuries that they were used, and used widely, in the Graeco-Roman world. First came the tension engine *oxybeles*, an enlarged gastraphetes crossbow on a frame. This machine inspired the basic two-armed arrow-thrower (I will call it a ballista here for consistency, though this name applies properly to later Roman arrow-throwers); it rested on a stock (*syrinx*), as we have seen, with a slider (*diostra*) and a winch system. The early engines had a linear ratchet attached to the stock, engaging pawls that were connected to the slider. The tension machine had a slider that was dovetailed onto the stock, and this feature may have been carried over to the first torsion engine (though it is absent in Fig. 2). There may have been some flexibility in the bow arms, as for a crossbow, in the earliest two-arm catapults, and so these machines might have been hybrid tension-torsion machines.

This early torsion engine was the creation of those Syracuse engineers in the 390 s BCE. Considerable and rapid development followed across the Greek world over the following century due, we can safely presume, to the weapon's success on the battlefield.

Catapult design became standardized by the 280 s BCE, to the extent that an engineer of the time wrote about a lengthy development process resulting in codification and rules of construction. Philon Mechanicus, one of the third-century engineers, who wrote widely about mechanical matters and who undoubtedly took part in the development of Greek torsion-powered projectile siege engines, emphasizes the incremental, empirical way in which progress was made: "*I suppose you are not unaware that the art [technē] contains something unintelligible and baffling to many people; at any rate, many who have undertaken the building of engines of the same size, using the same design, similar wood, and identical metal, without even changing its [i.e., the metal's] weight, have made some with long range and powerful impact, and others which fall short of the ones mentioned. Asked why this happened, they could not give the reason [aitia]. Hence the remark made by Polycleitus, the sculptor, is pertinent to what I am going to say. He maintained that excellence is achieved gradually through many numbers. Likewise, in this art [technē], since products are brought to completion through many numbers, those who deviate slightly in particular parts produce a large total error at the end. Therefore, I maintain that we must pay close attention when adapting the design of successful engines to a distinctive construction, especially when one wishes to do this while either increasing or diminishing the scale.*" (*Belopoeica* 49.12–50.12. See [5]).

Apart from the refined language, this account sounds like modern engineer-speak. Philon was aware of cumulative error, of the need to proceed carefully, varying parameters gradually, and recording results. This is empirical science at its best, and resulted in a flourishing of catapult designs by the mid 300s.

Following Dionysius of Syracuse, Phillip II of Macedon also sponsored and encouraged catapult engineers, and under him the size of arrow-throwers grew. By the time his son Alexander the Great went on his world tour (with a few friends) in the 340s he took with him stone-throwers (I will go with *lithobolos* for these Greek machine[6]) as well as "*many arrow-firers of different kinds.*" [6]. He is said to have used catapults differently—for covering fire as well as for siegecraft. It is reasonable to assume that his artillery helped him conquer five major cities and many smaller ones during the first five years of his bid to master the world. Projectile siege engines were changing warfare.[7, 8]

3.2 By the Numbers

The numbers that Philon discussed likely referred to the codification of torsion engine design and production. By a combination of practical tinkering with catapult performance plus a little mathematics, and proceeding carefully, the early engineers developed a spreadsheet approach for the use of catapult designers and for the craftsmen who built these engines. Thus the proportions of the different catapult components needed to be set just right for best performance; with these correct proportions, a ballista could be scaled up to make a more powerful machine. Given that such scaling up occurred widely over the centuries, for ballistas and all the other types of siege engines that we will discuss in this chapter, it behooves us to consider scaling here, to appreciate the achievement of the Greek engineers and to gain some insight into the intricacies involved—scaling up is not quite as straightforward as might naively be supposed. First though let us summarize the Greeks' codified proportions.

For a ballista, the arrow or bolt length should be nine times the diameter of the washers. Huh? Here is why. Catapult power came from the twisted skein: increasing the twist led to increased power, of course, but power also increased with the length of cord, in other words with the diameter of the skein. In Fig. 2 we have a cord that is twice as long as the distance between the washers—call this cord length $2h$. For a stronger spring, a longer cord of length nh is packed into the washers. The packing is made tight so that the bow arm—jammed through the middle—is held securely. Thus the spring strength is measured by the number n of cord lengths and so by the diameter of the washer. Stronger springs require wider washers.

[6] You may not be surprised to learn that there is disagreement among modern scholars concerning the naming and structure of these engines. Some say that lithobolos threw arrows as well as stones. Some say that the stone-thrower bow arms were in-swingers, rather than out-swingers like the standard configuration of Fig. 3. That is, the skein springs were widely separated, and the bow arms swung between them.

Thus a longer (and hence heavier) arrow requires more power to launch and so bigger spring washers. This 9:1 scaling rule for (early and relatively low-powered) ballistas tells us two things. First, as for bows the engine was matched to the size of the projectile—no surprise there. Second, and more interesting to archaeologists, we can estimate the power and size of an ancient catapult from its washers alone. Plenty of these bronze components have survived to the modern day, to be retrieved by archaeologists, unlike the wooden frames and bow arms. Those washers that have been retrieved show a clear typological sequence, indicating that many (perhaps seven) different sizes of catapult were manufactured.

What about the other ballista components? It seems that the bow arm length needed to be seven times the washer diameter, the spring frame height and width were set at $5\frac{1}{2}$ and $6\frac{1}{2}$ washer diameters respectively—the washer was the yardstick from which everything else followed.

For stone-throwers the scaling rule laid down by the Greek engineers was more complicated. The spring diameter (washer inside diameter) was related to stone projectile weight W via the formula $d_{washer} = 1.1(100W)^{1/3}$, in ancient Greek units (*minas* for weight and *dactyls* for length). Converting to metric units the formula becomes $d_{washer} = 2.1(229W)^{1/3}$, with d_{washer} in cm and W in kg.[7] We know that these statements were not just theoretical musings of ancient writers, that they were of practical value, because many carefully-rounded stone projectiles have been found (for example, 200 from a cache dating to 311 BCE at Salamis, in modern Cyprus) that were standardized by weight. Many other caches have been discovered by archaeologists, of many different weights (indicating many different lithobolos sizes); the most common have masses in the range $m = 5 - 15$ kg (call it 11–33 lb).[6]

We can see from the following geometrical argument that the development of scaling rules is a significant achievement, displaying some subtleties on the part of Mother Nature—meaning that the manner in which catapult size scales is not obvious, and requires detailed practical experimentation to unearth. Let us say that a neophyte Greek catapult engineer from the fourth-century BCE (Let's call him Bozon) tries to scale up an arrow-throwing version of the ballista of Fig. 2 by carefully magnifying the size of every component by a scale factor S. That is, the length and width and breadth of each bow arm was increased by a factor S, and same for the stock and spring frame. The spring washer diameter and depth increased by S, the length and thickness of cord increase by S, the arrow length increased by S and the bowstring thickness, frame size, universal joint, winch size...everything is supersized by the same factor. Would

[7] 1 kg of mass is equivalent to a weight of 2.2 lb.

the launch speed of the arrow be scaled similarly? If the launch speed of the projectile emerging from the original Mk 1 ballista was v_1, would the launch speed of the supersized Mk 2 projectile be Sv_1?

No. The Mk 2 ballista likely would break apart the first time it was fired. If the Mk 1 ballista was properly constructed then the spring cord was packed tight into the washers—and the same would be true of the Mk 2 engine. However the strength of a spring is proportional to its washer area (i.e., to the total thickness of cord inside) and so the tension in the bowstring would increase by a factor S^2. The thickened bowstring could withstand this force but the bow arms likely could not. The bending moment acting on each bow arm has increased by S^3 whereas its strength (its ability to resist snapping under a bending force) is proportional to cross-sectional area, which has increased by only S^2. So the bow arms likely would break.

For his Mk 3 ballista Bozon has used stronger wood, reinforced by iron straps and joints—this one won't break. He employs a larger crew to span the ballista, to winch back the bowstring, due to the increased forces, and places a scaled-up arrow in the groove, resting on the slider. It is launched ok, but the bow arms crash into the uprights of the spring frame, instead of being pulled up short by the bowstring as for the Mk 1 version—and the arrow launch speed is reduced, not increased.

Why? It's not obvious. When we get to ballista physics we will see that engine efficiency is better when the bow at rest is nearly flat, with the arms nearly perpendicular to the stock. For idealized bows, with rigid arms and bowstrings that do not stretch, the arms come to rest after the trigger has been released when the string is taut and straight. For such a configuration the bow arms are close to the spring frame uprights. In practice the bowstring is somewhat elastic and so, near the end of the internal ballistics trajectory just as the arrow is released, the bow arms will overshoot their equilibrium position due to inertia and bowstring stretching. This tendency will be enhanced for Bozon's supersized Mk 3 ballista due to the increased inertia of the bow arms. Due to overshoot, the arms may then slam into the spring frame.

The calculations of TA-6 tell us that arrow launch speed is given by, to a good approximation:

$$v_1 \approx \sqrt{\frac{2K}{m}\phi_0(\phi_0 + 2\alpha)}. \tag{1}$$

This equation applies for ballistas with bow rest angle near zero, for maximum efficiency. The bow arm angle is initially ϕ_0 when the string is pulled back to full draw length, and quickly reduces to zero during the firing phase, which lasts a fraction of a second. The angle α measures how much the springs are "pre-twisted" when the bow is at rest, with the string not drawn. Angles are not changed during the scaling-up process, because all lengths are changed by the same factor. In Eq. (1) m is projectile mass. Now we can see why arrow launch speed is reduced as the ballista is scaled up in size.

We can expect that the spring stiffness factor K scales to S^2K because stiffness is proportional to the area of the skein of cord, which is to say the area of the washer interior cross-section. Masses scale as S^3 assuming that material density is unchanged, and thus from Eq.(1) the Mk 1 launch speed of v_1 scales to the Mk 3 launch speed of $S^{-1/2}v_1$. Thus launch speed falls.

I do not know the Ancient Greek word for "bummer" but I can imagine Bozon uttering it at this point. He might reasonably if naively have guessed that launch speed would have scaled up to Sv_1 because the bow arm length does so, but reality in the form of geometry and engineering gets in the way.

Of course we can improve matters considerably by not scaling the projectile mass, or by increasing it a lesser amount. Bozon's Mk 4 ballista fires arrows of increased length but the same mass m as the Mk 1, which means that the Mk 4 launch speed is indeed Sv_1. Phew.

I have delved into the effects of scaling in some detail because it was an important factor for the development of all ancient Graeco-Roman projectile siege engines, and because it exercised the minds of the Greek engineers so much. Their rules for construction were the result. Now perhaps you can better appreciate the effort that they must have put in to get these rules. In the real world it is not obvious how machines scale and so to find out it is necessary to put in a lot of experimental design work, testing, tweaking parameters, and rebuilding and retesting. Just like modern engineering development but without the help of modern analytical tools and computers.

And yet there *was* a minor theoretical computation required by the Greek siege engine builders of the fourth-century BCE. Recall that the scaling rule for stone-throwers involved a cube root. It is generally not a trivial task, mathematically, to determine the cube root of a number, and at the time the scaling rule was introduced Greek mathematicians of the day had not solved the problem theoretically. It was solved by the Engineer Philon, whom I quoted earlier and who was one of many who were working on cube root extractions at the time; his method was the simplest and most practical for catapult construction. He wrote down a table with cube roots of relevant stone projectile weights, for the benefit of catapult craftsmen but also for modern historians: his writings show the range of stone projectile masses that were used in his day: 4.4–65.5 kg.[8] [9, 10].

3.3 Later Developments and the Roman Ballistas

Many of the early Greek engineers who developed the torsion arrow-throwers and then the larger stone-throwers are unknown to us; we know the names of a few who wrote about their achievements and those of others. The best known of them are listed here in chronological order from the fifth-century BCE to the second century CE, with a few scraps or fragments of their lives and achievements. Only one of the names is familiar today to the lay person.

Zopyrus of Tarentum played a significant role in developing the early tension engines (frame-mounted gastraphetes) in the late fifth-century BCE. Charon of Magnesia made major contributions to early (fourth-century BCE) lithobolos stone-throwers. Polydius of Thessaly was one of Philip II of Macedon's team of engineers, so likely worked on arrow-throwers.

We know a little bit more about Philon of Byzantium (a.k.a. Philon Mechanicus). He lived ca. 280–220 BCE mostly in Alexandria. Apart from his contribution to the cube extraction problem in mathematics and his technical writings on catapults, Philon provides insights into the working culture and mindsets of his contemporary engineers. He says that technicians in Alexandria were heavily subsidized by ambitious kings who fostered craftsmanship. Apart from engineering of catapults, he expressed concerns about their cost and the market demand for them, about their expediency and durability. (In fact there is some evidence that well-maintained catapults could remain in service for twenty years.) He advocated, based upon demand, for long-range catapults but not for one of the third-century developments, the repeating arrow-thrower, due to its short range. (Here is an interesting parallel to the roughly-contemporary repeating crossbow, half a world away in China. It, too, had limited power.) He also records that the status of engineers rose with the deployment and success of catapults.

Ctesibius was a contemporary of Philon and, like him, lived mostly in Alexandria.[9] A mathematician, inventor, and engineer, Ctesibius made contributions to many fields apart from siege engines, is known as the "father of

[8] This practical method of extracting cube roots is known to mathematics as "doubling the cube," or "the Delian problem," or "the Philon line." An alternative proof was (four centuries) later provided by another catapult engineer, Heron of Alexandria.

[9] Belatedly I should point out that Alexandria here refers of course to the large city in Ptolemaic Egypt, not to the thirty or so other cities that Alexander the Great founded or renamed after himself.

pneumatics" and invented the water clock. Perhaps surprisingly he is thought to have been very poor, unlike Philon. Yet another Alexandrian from this century, Dionysius (not to be confused with the tyrant of Syracuse, nor with a third-century CE pope) was the inventor of the (chain-driven) repeating catapult just mentioned.

Another third-century BCE engineer is famous these days for many things, but in his lifetime he was well known mostly for his contributions to military engineering. Archimedes of Syracuse is celebrated for his hydrostatics principle and his many contributions to mathematics. Indeed, in Europe his work on these matters has been written about and commented on since the Middle Ages. However it was his many military inventions—developed for and employed during the defense of his home city of Syracuse (213-212 BCE)—for which he was lauded at the time.[10] Such was his status that the commander of the besieging Roman forces ordered that his life be spared when they eventually succeeded in breaking into Syracuse. Despite this order a Roman soldier killed Archimedes, perhaps inadvertently. His death was regretted by Roman and Greek alike. Archimedes' siege weapons delayed significantly the Roman capture of Syracuse, a city with high sea walls. His engines were many and varied, and seem to have been designed for actions against Roman ships. Cranes and giant claws would grab and upend ships, to sink them. His giant stone-throwers—the largest in the classical world—could throw stone projectiles that weighed 3 talents (78 kg, over 170 lb). The range of these behemoths would have been short, but it is plausible that they could be employed from high ramparts against ships that were close by the walls. One less-plausible invention attributed to Archimedes is giant polished mirrors, said to focus rays from the hot Mediterranean sun on wooden ships and so set them aflame. Until recently historians were of the opinion that Archimedes held engineering in disdain, preferring purer intellectual pursuits, but this view is simply wrong.

Biton of Permagon lived in the third and second centuries BCE and wrote a treatise *Construction of War Machines and Catapults* (addressed to his sponsor, Attalus I, ruler of Pergamon in Anatolia). Heron of Alexandria takes us to the first-century CE and the end of classical Greece, which by this time had been absorbed into the Roman Empire. He developed the *carroballista*, a cart-mounted arrow-shooting ballista that was the most advanced torsion engine in the Roman army. It became standard equipment, so that each legion of the period was equipped with 55 carroballistas.

---

[10] According to Plutarch, Archimedes was spurred by king Hiero II of Syracuse into military engineering. Archimedes took an active part in directing the use of his engines during the siege, despite his age (mid-70 s).

Another late development by Greek engineers swapped out the sinew or horsehair springs for bronze springs. This new power source had the advantage of being more reliable and predictable, but it seems that the metal springs of antiquity were just not as powerful, so this variant was not produced in numbers.

Improved manufacture led to stronger engines for a given weight of projectile, and so it became possible to increase the angular excursion of the bow arms—the arc length that they would swing through before releasing the projectile. For centuries this angle had been about 35°, but then it increased (gradually?) to upward of 50°. This extra arc meant higher launch speed and greater range, and also of course greater stresses on the ballista frames.

These engineers knew of each other through their writings and met with their contemporaries across the Greek world. This international collaboration in research comes across as quite modern, though such technical teamwork in the Western world largely would be extinguished after the classical period for over a thousand years. Indeed, the important contributions made by these men (and they were all men—women played a subservient domestic role in Ancient Greece) were not fully appreciated by historians until the twentieth century, with analyses of their works by scholars who possessed an engineering or military background. The records are ambiguous as to how these engineers were perceived by the various Greek city-state societies around the Mediterranean in which they lived. Their status rose and their contributions were certainly appreciated by their military leaders, but perhaps not more widely. Different individuals and teams gained reputations for excellence in different fields (for example, Alexandrian and Rhodian engineers were widely considered to be the stone-thrower experts).

Vitruvius, a first-century BCE Roman writer on architecture and military technology, lamented the fact that his sources on technical matters were mostly Greek, and indeed much of Rome's artillery emerged from weapons or ideas captured from their enemies. This may be true generally—Romans were not as innovative as other peoples of classical antiquity, they adopted and adapted the inventions of others—at least this is the reputation that historians commonly attach to ancient Rome. Certainly many advances in engineering and technology were imported (shrunk-on iron tires from the Etruscans, catapults building design and metallurgy from the Etruscans, Carthaginians, and Greeks). In general Romans grudgingly held the view that Greek culture was in some ways superior to their own, even after their conquest of Greece (similar to the way some Americans used to think of their nation as an uncultured relative to their British antecedent).[11, 12]

Certainly Roman siege artillery in the second century BCE was captured from Carthaginians and Greeks. Engineers in the Roman Republic tinkered with foreign designs and improved upon them (it may have been a Roman idea to switch from a ratchet to a more compact gear mechanism for winding the engine; Romans added metal plates to strengthen frames, then introduced metal frames; they enlarged and more densely-packed the springs which increased their power and then added bronze casings to protect them). Metal frames made the Roman carroballista (which evolved from the Greek ballista, which it strongly resembled) more compact and also more accurate, due presumably to the frame rigidity. Julius Caesar, for example, writing about the siege of Avaricum in 52 BCE noted the accuracy with which his arrow-throwers could pick off individual Gaulish warriors. He used ship mounted arrow-throwers during his later contested landings on British soil. Both arrow- and stone-throwers became standard siege equipment. (The first-century CE Roman soldier and engineer Sextus Frontinus wrote: "The invention of [war machines] has long ago been completed and I don't see anything surpassing the state of the art." [3])

Late Roman arrow-throwing ballistas became very powerful. We know this from contemporary accounts and from a consideration of the physics. Thus from an anonymous fourth to fifth century CE pamphlet *De rebus bellicis* ("On the things of war") perhaps referring to the battle of Adrianople (378 CE, in the Balkans) we have the claim that one type of ballista could throw an arrow at least a kilometer: "*From this ballista, darts were projected not only in great number but also at a large size over a considerable distance, such as across the width of the Danube River*".[11] (The Danube is thought to have been at least a kilometer wide at that time and place.) From the late Greek scholar Procopius we have a description of ballista arrows/bolts: "*...and they place in the grooved shaft the arrow, which is about one half the length of the ordinary missiles which they shoot from bows, but about four times as wide*"[13]. Here he is describing ballistas that the great Byzantine general Belisarius placed on the walls of fortifications. He emphasizes the force with which such a bolt strikes its target. To a physicist the difference in dimensions of these bolts with conventional arrows is suggestive. A shaft that is half the length and four times the width will require a force that is a thousand times larger to cause it to buckle under the force of the bowstring[12] This in turn points to the power of the ballista. We will see later in this chapter that such ballistas as are described here (I take them to refer

[11] See the Wikipedia entry *Ballista* and references therein.

[12] The buckling force is called the *Euler critical load* and is proportional to r^4/l_{bolt}^2 where r is the bolt shaft radius and l_{bolt} is bolt length.

Fig. 3 The standard heavy artillery of the Roman army: a one-talent ballista. Public domain drawing of the BBC reconstruction, by Vissarion, modified by the author

to siege weapons of the Eastern Roman/early Byzantine empire) can throw an arrow/bolt over a thousand yards.

Ballistas became bigger and smaller. The *scorpion* was a compact one-man Roman arrow-thrower (the name referred to a different engine in the Greek world). At the other end of the scale we have the largest ballista in regular use loosing "one-talent" (26 kg or 57 lb) stone projectiles—see Fig. 3. (I will henceforth use *ballista* to describe both arrow- and stone-throwing Roman siege engines.) This latter engine had springs that were six feet high, stocks that were twenty feet long, and with a frame width of sixteen feet. (All these dimensions are a little smaller than the Greek equivalent, due to increased frame strength.) The Greek one-talent engine had an effective range of 160 m (530 ft) according to Philon, so presumably the Roman version fired its stone projectiles at least as far. Often, though, maximum range was not the most important feature of stone-throwers. Sometimes even the large ones were mounted atop a siege tower and brought up close to the enemy fortifications, where they launched heavy stones in flat trajectories (which would do more damage to walls than a long range, angled trajectory). Right until the end of the Western

Roman Empire, ballistas of one type or another were used for firing arrows or darts or bolts, but the stone-throwers were replaced in the late empire by a different type of catapult, as we will see.

Ballistas were an increasingly significant part of warfare for the Greeks and then for the Romans, for the better part of eight centuries. These engines threw various types of arrow, and threw stones both large and small. They conveyed status to the engineers who created them and to the crews who operated them. (Thus some Greek states included catapult shooting competitions along with gymnastics in their military training. Romans donated catapult washers as votive offerings to their gods, and portrayed catapults on their tombstones.) But how much of a difference did ballistas make to ancient warfare?

Quite a lot, according to some accounts: "*There is reason to associate the rise of large empires with the advent of the catapult.*" [10] This is no minor assertion. Another bold claim is that catapults changed societies as well as warfare[3]. The arrow-throwers were more accurate, more repeatable[13] and with longer ranges than bows or crossbows.[14] The stone-throwers were more powerful than any other weapon in history, and could knock down fortifications that were previously considered invulnerable. They conveyed a new status on engineers and crew, specialists who were at the forefront of the new semi-professional armies—so different from the massed levies of many ancient empires—so that technical knowledge challenged honor and virtue as the most desired characteristics of soldiers.

But there are contrary views among modern historians, who urge caution. There is confusion and misinformation among the ancient sources and the modern interpretation of these sources is consequently littered with misinterpretations and bad hypotheses. Campbell writes of the "*fragility of the accepted history of the catapult.*" [6] So what are we to make of this dichotomy of views? The scientist in me notes the improved performance of ballistas and concludes that they must have influenced the outcome of battles and sieges. If they were not so important as to change societies then they likely contributed to the outcome of wars (there were other reasons why the Greeks and then the Romans carved out large empires in Eurasia, but their possession of ballistas will have helped.[15]) I will return to this debate after discussing medieval siege engines, because the disagreements persist [8, 17].

[13] An early twentieth century arrow-throwing ballista reconstruction was sufficiently accurate that, when a number were shot at a target, one split an earlier arrow. [10, 14]. Range could be made consistent from one shot to the next by counting the number of clicks made by the winch during winding.

[14] "*The missile is launched with such force that it reaches not less than twice the range of a shot from a bow...*" Sixth-century CE writer Procopius War Against the Goths i.21.17 [15].

[15] I am reminded of Hilaire Belloc's words from the nineteenth century:"*Whatever happens we have got, the Maxim gun and they have not.*"[16] It was ever thus.

3.4 Ballista Ballistics

The idea here, as always, is to provide a *physical* model of the dynamics of our projectile weapon. This means that I make a (simplified) approximation to the shape of the weapon—here to the bow arm length, frame dimensions, angles, etc.—and derive an equation of motion to describe the dynamics. This procedure can accurately estimate parameters of interest, and in particular can tell us to a pretty good approximation what the projectile launch speed will be. What it does not do, however—and this is important—is convey the engineering intricacies of the weapon. There are a lot of moving parts that have been devised and constructed with care, for these siege engine weapons, more so than for the weapons of earlier chapters. Readers may appreciate the sophistication of the process of making a composite bow and yet get the wrong impression about the sophistication of ballistas and their brethren. We are entering the age of the machine, with these siege engines, and the devil is in the engineering details. My physical model might give you the wrong impression that there are only a few moving parts that need to be accounted for. While this is a reasonable approximation concerning the engine dynamics, there is a danger that you may come away from my account thinking that ballistas are just another projectile weapon—a bow on a frame and on steroids.

If you are not so far impressed by ballistas as a giant technological step forward, I urge you to peruse an account of ballista reconstruction, such as that of Wilkins, who built a ballista based on the detailed plans written down by the Roman engineer Vitruvius.[11] This account and others convey the engineering complexity of these engines.

A derivation of Eq. (1)—of the physics of ballistas—is provided in TA-6. I make the same assumption as for bows—that the ballista geometry can be well approximated by that of a spring bow with stiff limbs. Again I assume that the bowstring does not stretch. Also as before I surmise that friction is not an important factor during the internal ballistics phase of motion—this is perhaps a little less true than for bows due to the large forces that act within the skein springs. Given little friction it follows that energy is more-or-less conserved during the launch of a projectile, and that the heat and sound generated have insignificant energy compared to that possessed by the springs or bow arms or projectile. Also as before, I take it that the bowstring weighs much less than the arms or projectile.

Even with all these assumptions we can reasonably expect that Eq. (1) provides an estimate of projectile launch speed that is a fairly good approximation to that of a real ballista with the same dimensions. In our model the energy that is initially stored in the springs becomes divided so that, at different times during the launch phase, it can be found in the bow arms and projectile as well as in the springs. As for bows we find that ballistas are very efficient—the bow arms angular speed goes to zero as the bowstring straightens up, and the arms give up their energy to the projectile (see Fig. 8 of Chap. 4). Because of pre-stressing, the potential energy stored in the springs never goes to zero, and so bow efficiency never becomes 100%. In TA-6 ballista efficiency is found to be (for the spring-bow model with bow arm angle ϕ_1 as close to zero as is physically possible[16]):

$$\epsilon = 1 - \left(\frac{\alpha}{\phi_0 + \alpha} \right)^2 . \tag{2}$$

From this equation we see that more pre-stressed engines are less efficient. They generate higher launch speeds as we saw from Eq. (1), but operate less efficiently. Recall that the braced angle ϕ_0 increased over the centuries, from about 35° to about 50°, which Eq. (2) tells us that ballistas became more efficient as designs changed.

I wonder how many readers noticed that neither the estimated ballista projectile launch speed v_1 of Eq. (1) nor the ballista efficiency ϵ of Eq. (2) depend upon ballista dimensions. I imagine those who are engineers did notice. If it is exact, Eq. (2) is saying that efficiency is the same whatever the scale of the ballista. If Eq. (1) is exact, it tells us that launch speed does not depend on bow arm length l, bow string length L, or stock length Y. Yet the Greek engineers found empirically that there was an optimum choice for these lengths, which varied with washer diameter (which is to say, with spring strength). Why the difference?

The assumptions made in TA-6 deriving Eq. (1) are too simple—they make the calculations tractable but are not especially realistic, when examined closely. They are reasonable approximations to the real world, but can be expected only to yield approximate results. Thus for example spring force was assumed to be proportional to the angle of spring deflection. This is almost right, as Landels has shown[4], but is not quite right. Similarly in TA-6 I assumed that the bow arms were perfectly stiff and the bowstring was inextensible. Back on planet Earth there is no such material—but the bendiness of thick wood bow arms is small, and the stretchiness of hemp bowstring is small. Bottom line: the real-world empirical engineering of the Greek engineers spotted a (slight) dependence on ballista parameters that my simplified analysis couldn't see.

---

16 From Fig. 3 you can see that it might not be possible for ϕ_1 to be zero—for the bow arms to be parallel—because the bowstring would foul the spring frames.

All these disclaimers do not mean that the predictions of Eq. (1) should be discarded, I hasten to add, but should be viewed as approximations. If Philon tells you that the range of a specified ballista is 90 m whereas Eq. (1) says 80 m, believe Philon. But if Bozon tells you the range of his Mk 7 is 700 m and Eq. (1) says 300 m, believe Eq. (1).

Segue to historical claims about ballista ranges.

3.5 Two Benchmark Ballistas

I have built (mathematically) two ballistas that I will load with stone projectiles of various weights and then release the trigger (mathematically) to see what kind of ranges they attain. There are several practical and curious—I will let you decide in which sense—people who have built real ballistas, not just mathematical ones, and tested them. Their efforts and motivations will form part of the last chapter, in an overview of experimental archaeology, a *bone fide* and fun branch of the field. Here I will briefly report on the results of two early twentieth-century researchers—very different people, but with a common passion—who made small ballistas and reported on how they performed.

(Half) my two mathematical ballistas are shown in Fig. 4, fully spanned/drawn and (dashed lines) undrawn. The big one has bow arm length $l = 1.5$ m (5 ft) and a draw angular range of $\phi_0 - \phi_1 = 50°$; the small one has $l = 0.6$ m (2 ft) and $\phi_0 - \phi_1 = 35°$. The drawings are to scale. The big ballista has a draw weight of $F = 2500$ lb, and for the small ballista $F = 500$ lb. The small ballista is pre-stressed only slightly, $\alpha = 5°$, for efficiency whereas the large one is more pre-stressed, $\alpha = 30°$, for power. How well do they work?

They are efficient machines—they can thank their gastraphetes ancestor for that. The small (large) ballista efficiency is 91% (82%). For spherical stone projectiles of mass m the maximum achievable ranges of these two engines are plotted in Fig. 5. The launch speeds are determined by the analysis of TA-6 (not quite the same as Eq.(1) because the bow arm rest angles ϕ_1 are not zero). For these launch speeds, maximum range is found from the analysis of TA-1, which includes the effects of aerodynamic drag—considerable effects, given the high initial speeds (v_1 exceeding 100 ms^{-1} for the big ballista throwing light projectiles). To make these calculations I have assumed that the stones are of average density (2500 kg m^{-3}), then calculated their size and cross-sectional area for each projectile mass, and so determined the drag factor b. (Remember that there is no backspin for ballista projectiles, and so no aerodynamics lift—just drag.)

We see that the small ballista can throw a 1-lb (0.454 kg) stone 244 m, though it struggles with heavier projectiles—this small ballista is little more

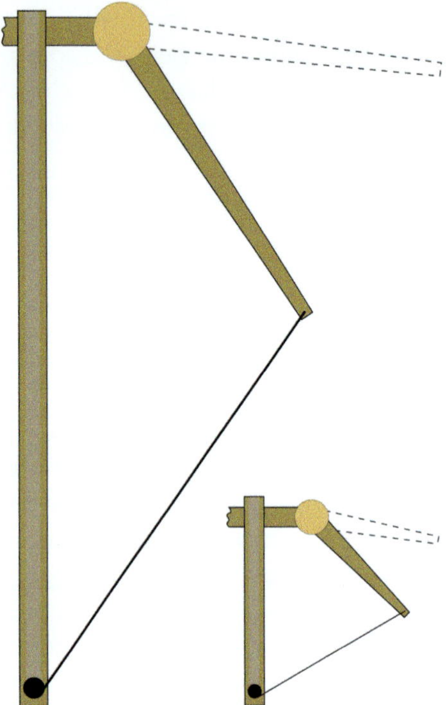

Fig. 4 Large and small ballistas, for the purposes of mathematical simulation. The dimensions are all to scale (except the bow and frame thickness—too slender for real ballistas)

than a gastraphetes crossbow in terms of draw weight, but would be an effective range weapon against infantry. The big ballista is a different beast. It can cast a 2-kg stone to over 700 m and (off the graph) a 0.454-kg projectile a kilometer (that is, it sends a 1-lb stone over 3200 ft). A one-talent[17] or 26-kg stone projectile can be thrown about 155 m. My big ballista does almost as well as the ones that Philon has described.

Let us look at the performance of some reconstructed ballistas and compare with Fig. 5. A century ago Schramm[14] and Payne-Galway [17] independently built and tested small ballistas, and both these machines were able to throw a one-pound stone a distance of 300 meters. Our predictions for the small ballista of Fig. 4 are a little short of that, but similar. The three ballistas (the two real ones from the early twentieth century and my mathematical model) all have slightly different parameters, but the sizes are similar. From this

[17] Profuse apologies to math-averse readers for using so many different units here (Metric, English, and Ancient Greek), but it goes with the territory.

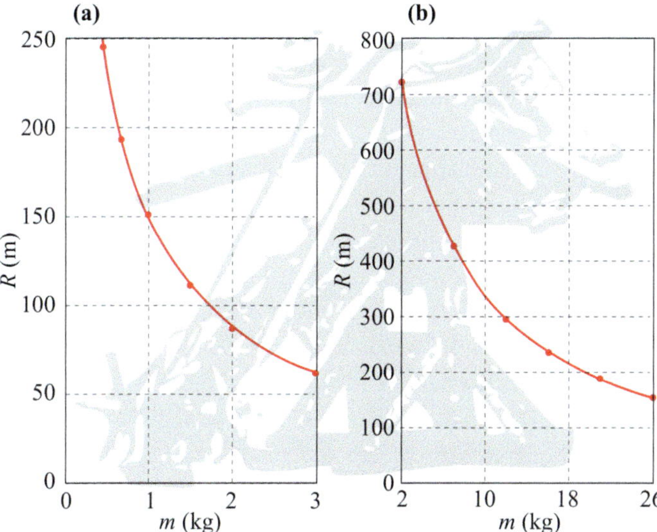

Fig. 5 Simulation results. **a** Small ballista range R versus projectile mass m. **b** R versus m for the big ballista

comparison I conclude that the simple model of TA-6 makes pretty reasonable predictions for ballista launch speeds, from which the standard aerodynamics calculations of TA-1 turn into range predictions.

The only large ballista reconstruction for which I have any details is that made by the BBC (British Broadcasting Corporation) in 2002 (see Fig. 3) [18]. This impressive machine launched a 1-talent stone a distance of around 90 m; on its second throw a bow arm snapped. (Philon would have expected this—he had written twenty-three hundred years earlier that bow arms breaking was the most common malfunction of ballistas.) My mathematical big ballista shows (Fig. 5b) that Philon's claim is physically possible assuming a draw weight of about 2,500 lb, so we can conclude—unsurprisingly—that ancient Greek engineers were better at building ballistas than is the BBC.

4 Summary

In this chapter we have learned about the historical development and performance of the two-armed vertical-axis siege engine. The history is confused for technical reasons and this confusion extends to the modern name given to it; I have generally called it a ballista whether it threw arrows or stones. The early version developed from the Greek gastraphetes crossbow and so threw arrows. This early machine grew in size and sophistication due to research programs in the fourth-century BCE Greek Mediterranean world. Later the Romans would take over both ballista development and Greece. There were several different forms of the machine, from small arrow-throwers to very large stone throwers, and these many forms acted as the artillery of the Greeks and Romans for eight centuries. The arrow-throwers were very accurate; they were light and mobile and could be used on open battlefields, but also during sieges. The stone-throwers were heavier and not mobile—they were set in place and used for sieges, by both attacking and defending forces.

"*The large engines deployed by the Romans at Yodefat were clearly intended to demolish the walls as well as kill defenders inside the town.*" (Holley[20]) Yodefat was a fortified town on the Golan Heights which was subjected to a bloody, 47-day siege by the Romans during the First Jewish War, 66-73 CE. The besiegers used 160 engines to besiege the town, with arrow-throwers picking off defenders and stone-throwers used to demolish ramparts. Just how effective siege engines were against defensive walls is a question for the next chapter; here we see that they were useful in that role even in the first-century CE. They could hurl a 26-kg stone perhaps 100 m, which placed the crew within

Fig. 6 Reconstructed arrow-throwing ballista at Gamla, another town besieged by Vespasian during the First Jewish War.

range of defenders' arrows. Arrow-throwers (Fig. 6) were out of range of enemy archers, though not, of course, of enemy arrow-throwing ballistas.

Once the designs had been perfected, ballistas were complex machines with many components. The two arms were powered independently by springs, each consisting of a skein of horsehair or sinew cords. These had to be balanced and carefully maintained. Crews could be trained easily, relative to other projectile weapons such as bows and slings. ballistas were mass produced in the early Roman Empire and formed an important component of each legion.

References

1. M. Landrus, *Leonardo da Vinci's Giant Crossbow* (Springer Science & Business Media, 2010)
2. U. Singh, *A History of Ancient and Early Medieval India: From the Stone Age to the 12th Century* (Pearson Education India, Delhi, 2008), p.272
3. S. Cuomo, The sinews of war: ancient catapults. Science **303**, 771–772 (2004)
4. J.G. Landels, *Engineering in the Ancient World* (Constable, London, 1978)
5. M.J. Schiefsky, Technē and Method in Ancient Artillery Construction: the Belopoeica of Philo of Byzantium, in *The Frontiers of Ancient Science: Essays in Honor of Heinrich von Staden* (De Gruyter, Berlin, 2015). https://doi.org/10.1515/9783110336337
6. D.B. Campbell, *Greek and Roman Artillery 399BC - AD 363* (Osprey, Oxford, 2003)
7. E.W. Marsden, *Greek and Roman Artillery* Historical Development. (Oxford University Press, Oxford, 1969)
8. C.W.C. Oman, *The Art of War in the Middle Ages* (Cornell University Press, Ithaca, NY, 1953)
9. D.B. Campbell, Ancient catapults: some hypotheses reexamined. Hesperia **80**, 677–700 (2011)
10. W. Soedel, V. Foley, Ancient catapults. Sci. Am. **240**, 150–161 (1979)
11. See the Roman Military Research Society's detailed online article *A suggested reconstruction of Vitruvius' stone-thrower* by A. Wilkins at http://www.romanarmy.net/pdf/Reconstruction%20of%20Vitruvius%27%20Ballista%20Ver%201-05.pdf
12. E.W. Marsden, *Greek and Roman Artillery* Technical Treatises. (Oxford University Press, Oxford, 1971)
13. Procopius *Gothic war* chapter XXI. Available online at https://www.gutenberg.org/files/20298/20298-h/20298-h.htm#FNanchor_103_103
14. E. Schramm, *Die antiken Geschutze der Saalburg, 1918* (Reprint, Bad Homburg, 1980)
15. A. Wilkins, *Roman Imperial Artillery: Outranging the enemies of the empire* (Archaeopress, Oxford, 2024)

16. H. Belloc *The Modern Traveller* (1898). Available online at https://www.gutenberg.org/files/61521/61521-h/61521-h.htm
17. R. Payne-Gallwey, *The Crossbow* (Holland Press, London, 1903)
18. G. Macdonald, G. Mullen, A. Wilkins, Building the BBC Ballista. Timber Framing **65**, 10–19 (2002)
19. W.S. Smith, W. Wayte and G.E. Marindin, Ed.s, “A Dictionary of Greek and Roman Antiquities," entry *Tormentum* (John Murray, London, 1890). Available online at https://www.perseus.tufts.edu/hopper/text.jsp?doc=Perseus:text:1999.04.0063
20. A.E. Holley “Stone projectiles and the use of artillery in the siege of Gamla," *Gamla III: The Shmarya Guttman excavations 1976-1989, finds and studies: part 1*, 35-56. Israel Antiquities Authority, Jerusalem (2014)

7

Horizontal-Axis Siege Engines

Chapter Summary The double-pendulum aspect of staff-sling dynamics is shown to enhance its performance. The onagers, with and without sling, are analyzed and their performances predicted. The roles of these and other horizontal-axis engines in ancient siege warfare are discussed. Onagers are compared with ballistas, and the eventual dominance of onagers in the late Roman Army is described. An explanation of the general observation that siege engines were more effective at short range is provided. The history and significance of traction trebuchets are presented. Two "benchmark" traction trebuchets are analyzed and their performances (launch speed, range, efficiency) predicted and compared (with each other and with the performance capabilities claimed of ancient machines). It is shown for traction machines that the many and various engine parameters need to be chosen carefully to optimize performance. Counterweight trebuchets are introduced and their origins speculated upon. The performances of two varieties of CW trebuchets (fixed CW and swinging CW) are derived and compared with each other and with the traction trebuchet. The historical spread and evolution of trebuchets is given, and their effectiveness as the most powerful mechanical war machines is emphasized.

"*The trebuchet dominated warfare far longer than any other form of artillery, yet it remains the least understood piece of military ordnance. It holds the distinction of being the most powerful form of mechanical artillery ever devised.*"–(Paul Chevedden, American historian, 2000) [1].

M. Denny, *Slings and Arrows*,
https://doi.org/10.1007/978-3-032-08563-4_7

1 Engines of Destruction

This is the final historical chapter, introducing as it does the last class of ancient projectile weapon that we will investigate together. Horizontal-axis siege engines include the Onager and traction trebuchet of classical antiquity, and the gravity-powered machines of the Middle Ages. The latter were the largest of all the siege engines, and we will see that there developed something of an arms race at this time: bigger siege engines could breach bigger castle walls and so castle walls got bigger, and so siege engines got bigger. The main point for me (physicist/engineer/nerd) is that these large siege engines of the eleventh to fifteenth centuries represent, not exactly a break with the past, but certainly a sign of things to come.

I would like to spend a few paragraphs outlining my take on the broad sweep of history that has formed the backdrop to the weapons we have so far encountered, to persuade you that the trebuchets of the late medieval period represent the first signs of a sea change in the lives of mankind. (They did not *cause* this sea change; they are the earliest example I am familiar with that *portray* it.) My background is physics and mathematics, not history or archaeology, so perhaps the eyes of a professional historian will roll at my conclusions—or perhaps not. I put them out there anyway. I think that they hold up to scrutiny and that you will find them interesting—just bear in mind where I am coming from.

Forty thousand years ago, when atlatls (and perhaps boomerangs, bows, and slings) first were manufactured, your ancestors and mine wandered the Earth as hunter-gatherers, alongside other hominid species who soon (or not so soon) left the scene. Then came dogs and chickens and goats and horses and other animals that *Homo sapiens* domesticated: most of humanity (though by no means all) became nomadic pastoralists, wandering the landmasses of our world ever looking for new grazing land and berries. Then came wheat and rice and other plants that we domesticated (cultivated and improved as food sources—early genetic engineering). An increasing and significant fraction of humanity adopted the agrarian lifestyle—they settled and built villages, then towns and cities. Compared with the nomadic peoples—who were still out there—agrarian populations grew fast. This rapid growth, combined with their high population density, made them relatively vulnerable to epidemics and so, while the number of agrarians grew, their average lifespan likely fell. Their diet was different; nomads ate mostly meat. I wonder if, had they been geographically isolated, these two groups of people would have diverged into different species.

But of course they weren't isolated from each other. As we have seen, they fought and intermingled.

Both groups invented things that were useful for their own needs—such as stirrups for the nomads, plows for the agrarians. Probably the agrarian societies invented more things, because their numbers were higher. Say that one ancient society has made a number N of inventions that are useful to them—these inventions might be gadgets, or ideas about agriculture or animal husbandry, or boat building or horse breeding. Because our ancestors were smart (at least compared with other species) they tinkered with their inventions. Innovation, adaptation, and incremental improvements in technology thus developed. Let us say that, for the society with N inventions, each century they made a small number δN of improvements in these inventions (an improved plow, better conditions for growing barley or managing pigs, a superior way of rigging boat sails, a refined stirrup).[1]

At first, in ancient times when N was small and δN was next to zero for all the groups of people in the world, technology advanced slowly. The sword or bow or housebrick they used were pretty much identical to the swords and bows and housebricks of their ancestors a millennium earlier. Tinkering and incremental improvements in those days would likely have been in isolation. Thus, a potter at some distant place and time happened upon a slightly different way of firing pots, and it led to better pots. So they retained the idea, and it spread around their clan or city, and so pots evolved—*but only pots.* The pot-firing tinkering led to no improvement in other things.

But as the number N of invented things increased over centuries, it became more likely that an improvement or refinement in one would have consequences for other items of technology.[2] At some point in time—not long after the heyday of the gravity-powered siege engines that we will soon investigate—the number of beneficial increment δN grew in proportion as the number of inventions N grew.

Now, any physicist or mathematician who is told that a change δN increases with increasing N will immediately exclaim "exponential growth!" Inventiveness feeds upon itself, from this time on. The exponential growth in technology, and the techniques for developing technology, grew slowly and then quickly,

[1] E.g., plows improved with the invention of the detachable plowshare about three millennia ago in the Middle East [2]; barley yields have improved by selective breeding for larger seeds; pig management strategy (a change in breeding regime) improved in N. Europe about six millennia ago [3]; the lateen sail was introduced in the second century CE in the E. Mediterranean; in the same century paired stirrups were introduced in China [4].

[2] E.g., In the Middle Ages the S. European carvel boat hull technology met the N. European central stern rudder, resulting in a new type of ship (the carrack) which was much better suited for the open ocean than were any existing European ships with only one of the two innovations [5].

resulting in the Industrial Revolution (from the middle of the eighteenth century) and the creation of the modern world.

Perhaps this is fanciful but, it seems to me, with the benefit of hindsight we can see the seeds of this growth in inventiveness—the very first hints of technology booting itself up—in the arms race between siege engines and castles.

2 Staff Sling

The most basic, earliest, and least controversial of our devices, the staff sling (Latin *fustibalus*) is simply a sling on a stick. It seems too humble to be counted as a siege engine, and yet it was used as such. It is a one person, hand-held implement, employed unchanged for thousands of years. It was used as a siege engine due to its long range and high trajectory, throwing stones or incendiaries over walls. The staff sling is the end of the line for solitary throwers: all our remaining siege engine projectile weapons will be (mostly or entirely) crewed by more than one person.

The staff sling is a straightforward extension of the sling—literally. It must have occurred to many sling users in the Stone Age that their weapon would be more effective, with a greater projectile speed and range, if they had a longer throwing arm. What to do? Attach the retention cord of their sling to the end of a straight stick, loop the end of the release cord over the same end, and *voilá*—the first staff sling. Some experimenting with the release cord length and with the loop release mechanism will have been required, but it will not have taken long to perfect. Staff slings turned out to be easy to use, unlike their sling antecedents. This ease of use explains the persistence of staff slings well into the second millennium—no need for years of practicing from childhood; raw recruits in an army could be taught the use of this weapon quickly.

The simple process of adding a stick changes the sling's characteristics and use. Thus, the staff sling took two hands to operate effectively, and so the thrower, if a soldier, could not simultaneously hold a shield. Two arms meant more power, and so heavier projectiles could be thrown. Two arms also naturally meant a vertical throwing action—the plane of acceleration of the staff-sling bullet is vertical. This was pretty much the only way to use a staff sling—there were no variations in style, in the manner in which a thrower could hurl a projectile. The staff was held horizontally overhead and behind the slinger, with the sling dangling its projectile in a (large, compared with a regular sling) pouch. The slinger then accelerated the staff upward through a quarter turn, or so, and the projectile was released in a high, arcing trajectory.

The throwing style explains why the staff sling was simple to use and why the projectile trajectory was high. Muscle power limited the staff length to at most two meters, call it $6\frac{1}{2}$ feet. (The Chinese staff sling—the *piao shi* or "whirlwind stone"—at its peak popularity in the Song Dynasty army for siege defense, was standardized at about $5\frac{1}{2}$ feet.) There was no need for the thrower to worry about aligning the bullet spin direction with its velocity direction (another reason why it was easy to learn) because there was no spin stabilization—the spin direction was perpendicular to the direction of travel, as for a golf ball. The projectile had backspin, and so perhaps may have generated some aerodynamic lift when in flight, thus extending its range a little. Because of this feature it is likely that staff-sling bullets were shaped to be spherical, as much as possible, to avoid tumbling which led to lack of accuracy and reduced range.

We can see quite easily why the range of a staff sling is long—longer than that of a sling for the same spherical bullet. First, the sling extended the thrower's arm length, and so there was a lever effect (think atlatl). Then there was the backspin and finally, there was the sling. The double-pendulum effect that we saw operating in the sling works in spades for the staff sling. In practice the staff sling was not generally used to extend range but rather to project heavier bullets. The high trajectory meant that accuracy was poor at short range, and so we arrive at the conclusion (from simple considerations of the internal ballistics) that the staff sling was useful for lobbing heavy projectiles to the kind of ranges that were attainable by slings and bows.

Historically, this seems to be the case. Thus in the Late Roman army: "*The archers and slingers set up bundles of twigs or straw for marks, and generally strike them with arrows and with stones from the fustibalus at the distance of six hundred feet.*" So says Vegetius in *De Re Militari* ("The Military Institutions of the Romans"), in a section titled "The Drilling of the Troops." There are three implications of this snippet about fourth-century Roman army training. First, they did use the staff sling. Second, its use was taught to raw recruits, and so must have been easy to master, and quite quickly. Third, newbies were expected to hit moderate-sized targets at two hundred yards.

Given the ease of use and rate of fire, the economy of manufacture, the weight and variety of projectile, and the ranges attainable, it is easy to see why staff slings found a place in many ancient and medieval armies around the world. Thus ancient Greeks, Carthaginians, Romans, and Vandals certainly used them, as did early medieval Byzantines and Chinese and late medieval Europeans. Staff slings were used especially for siege work on both sides of the walls, and were used on board ships until well into the seventeenth century [6–8] (lobbing incendiaries at enemy warships, for example). In the gunpowder age they were retained for a time as grenade launchers.

We need not bother ourselves with any calculations of staff sling maximum range. We know already for a regular sling that range can exceed a quarter mile, and so Vegetius' distance of less than half that will certainly be attainable with a sling that has been extended, i.e., with a longer lever arm. However it will be worthwhile to report here the results of an analysis of staff-sling action to demonstrate the kind of advantage to be gained from the double-pendulum effect. This effect, already seen in the rather complex sling calculations of Chap. 2, is revealed more transparently here where there are fewer competing effects, such is the simplicity of staff-sling internal ballistics. Also the double-pendulum effect of slings will be an important part of traction and counterweight trebuchet motion, to be considered later. For these machines the slings were quite long, which greatly complicates both the analysis and operation of these engines while doubling down on the performance benefits.

Consider a staff sling held horizontally, with the sling and projectile dangling vertically. The slinger flicks the staff until it is vertical, like an angler casting a rod to propel a lure into a river. Let us say, to keep things simple, that she flicks the rod at constant angular speed; we can calculate the launch speed v_1 of the projectile, given staff and sling lengths a and b respectively (see TA-7). For comparison we can calculate the launch speed for a projectile on the end of a staff of length $a + b$, with no sling—let us call this the "lever effect" launch speed, v_0. The ratio of the two launch speeds is plotted in Fig. 1. Note that

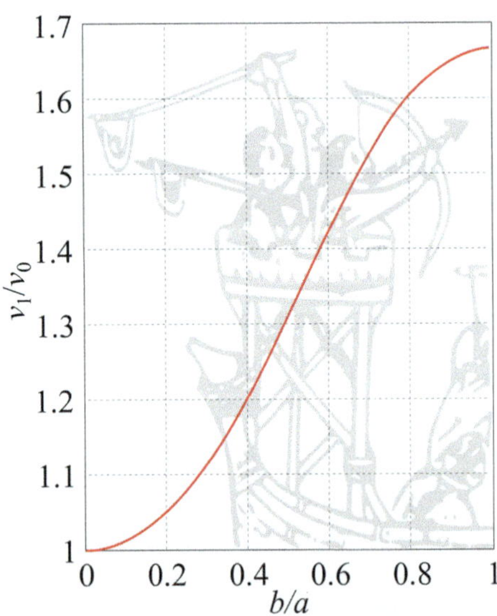

Fig. 1 Double-pendulum benefit (gain in launch speed) versus ratio of sling length to staff length

the staff-sling speed is higher, even though the lever effect is the same for both; this tells us that Fig. 1 is showing us the double-pendulum benefit alone—it boosts the launch speed by over 50%.[3]

There are in fact *two* benefits for a mechanical projectile weapon to make use of a sling. First, as found here, a sling boosts projectile speed. Second, as we will see for our larger siege engines, it provides a trigger. For the staff sling its trigger is a simple loop on the release cord, which hooks over the end of the staff, and releases at some desired angle (here when the sling and staff form a straight line, say, to maximize the lever effect). The angle is determined by the shape of the staff end over which the sling is looped. For other siege engines the loop is attached to a hook that is carefully angle so that the retention cord naturally slips off it at the desired angle. For some engines—those with short slings, such as the traction trebuchet—this is the main reason for employing a sling. For others, such as the counterweight trebuchet, the sling is very long and serves both purposes [9–11].

3 Onager

It seems that in 210 BCE Philon made reference to a one-armed stone-throwing machine, and this is the earliest reference to the weapon that we know as the *onager*. Often described as a mechanized staff sling, the onager's source of power is a single torsion spring (of the ballista type, made from a skein of sinew or horsehair) with a horizontal axis, which applies torque to a single throwing arm which rotates in a vertical plane—see Fig. 2. It seems natural enough for the Greeks to have invented a one-armed torsion engine before the two-armed ballista; if so then they must have rejected the one-armed machine as inferior, with good reason. We hear little about such machines from the Greeks after their invention of the lithobolos, and little enough from the Republican and early imperial Romans, who adopted the two-armed ballista, as we have seen.

So why was the onager inferior to its two-armed cousin and why, in that case, did the Romans adopt it as their main artillery weapon in the fourth century, supplanting the ballista? These are the questions that we will explore in this section, before turning to an analysis of this odd machine [12–15].

There are plenty of records attesting to the surprising switch, by the Romans, to such a relatively crude machine after at least five hundred years of contin-

[3] Weasel words: I have assumed, to make the calculations of TA-7 more transparent, that the staff was flicked at constant angular speed; in the real world it is likely the slinger would accelerate the staff. However this would only serve to increase the double-pendulum effect further, and not reduce it. My simpler calculation establishes the point that slings boost launch speed.

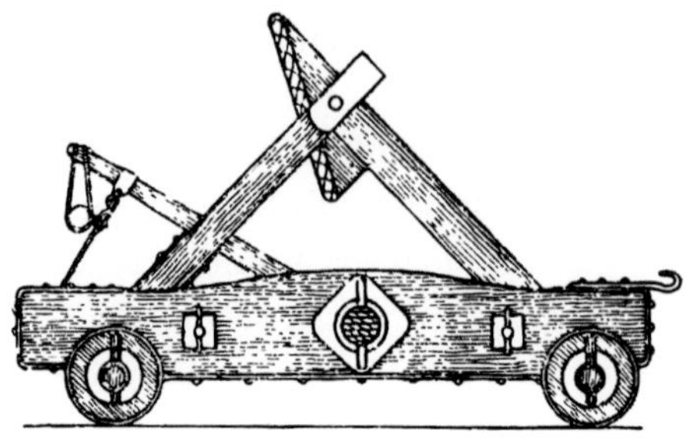

Fig. 2 Two onagers of the later type, with short slings. Public domain images from *Antique Technology* by H.A. Diels (1848–1922)

uous ballista use and intermittent improvement. The onager is mentioned by Marcellinus in 353 CE who says that the recoil (when the arm hits the stop or buffer—see Fig. 2) is so great that a large version of this siege engine, requiring a crew of eight to winch back the arm, could not be placed atop the walls of fortifications because the recoil could damage the stonework. This infamous kick is what earned the weapon its name: an *onager* is a type of Syrian wild ass, which has a vicious kick. Marcellinus also refers to the onager as a "scorpion," presumably because the vertical action is reminiscent of a scorpion raising its tail to sting. However the same name had already been used to describe an earlier torsion engine (a small dart-throwing ballista) and thus served to sow confusion in the minds of already-confused historians. The terminological uncertainties regarding projectile siege weapons continue, you see, beyond classical antiquity and will be with us to the end.

The inferiority of the onager compared with its much more sophisticated two-armed cousin is due in part, of course, to the kick. The kick meant that the onager was inefficient, as we will see, because energy was transferred to the buffer rather than to the projectile. The kick meant that the onager needed to be built sturdily, and so was heavy. It also meant that there could be no universal joint, due not to the weight but rather to the shock caused by the kick. Thus aiming the weapon consistently was more difficult, and so it was less accurate than a ballista. The main source of breakage of onagers was, as you might expect, the arm thwacking against the buffer.[4]

It is probable that the onager had been in use by both Greeks and Romans since Philon's time, but was overtaken if not replaced by the ballista. Historians

[4] Trial and error will have led to Roman engineers building onagers with the buffer placed so that the point of contact on the arm was two-thirds of the way up from the pivot. This is the so-called *center of percussion* and an onager arm (or sword or baseball bat) struck sharply at this point will reverberate less than if struck at any other point along its length.

disagree on this point. Some maintain that the two engines were used alongside one another from the heyday of the Western Empire right until its end and beyond, with both falling out of favor when the traction trebuchet reached Europe.[5] In this view, the ballista was in use until medieval times. Others say that the surviving evidence is confusing and that onagers more-or-less replaced ballistas in the Roman Army sometime before the fifth century [16–23].

Whenever it happened, the onager seems to have become more common than the ballista over the early centuries of the first millennium. Given the onager's shortcomings, why did this happen? (Vegetius in the late fourth-century CE stipulated that legions should have one onager per cohort—ten per legion[6]). They were simple to build and use. There was only one spring, and so it was not necessary to make complex components to effect the balancing of two springs. They were relatively robust. They scaled up easily (of which more later) and so became large. The consensus of historians seems to be that, by the fourth century, the Roman Empire was failing administratively and as a consequence was unable to maintain the manufacturing and technical infrastructure to make ballistas, or to train their crews. The onager was altogether simpler to make, master, and maintain.

Thus Old One-Arm was metaphorically brought out of retirement, dusted off, and put in the front line. Originally without a sling, the onager was used for disrupting enemy lines and destroying enemy fortifications. Once the (short) sling was added it became more powerful, as we will see (recall the staff-sling double-pendulum effect). Having survived in memory for several hundred years, there was a late blossoming—an Indian summer—that lasted from the fourth to sixth centuries CE (longer, according to some historians) surviving the death of the Western Empire and continuing to serve in Eastern (Byzantine) Empire armies [24, 25].

3.1 Buckets

The early fourth-century Roman onager had a bucket or indentation at the end of the throwing arm, instead of the sling shown in Fig. 2; it was more like a mechanized stone-throwing atlatl than a mechanized staff sling. The arm is often portrayed as a spoon, though there is no evidence to support this construction.[7] This early engine is particularly easy to analyze—few moving

[5] D.B. Campbell, personal communication, June 2025.

[6] *De re Militari* IV–22.

[7] Indeed there is very little archaeological evidence for the onager at all. It had no characteristic metal parts analogous to the ballista washer that could survive over millennia, and was composed almost entirely of perishable organic matter.

parts—and in TA-7 you will find a derivation of the following equation for bucket-onager launch speed:

$$v_1 \approx \sqrt{\frac{K(\theta_0^2 - \theta_1^2)}{m + \frac{1}{3}M}}. \tag{1}$$

Here θ_0 is the initial angle of the arm, measured relative to the vertical, when fully drawn back and ready for release. θ_1 is the (smaller) angle of the buffer. Note that the launch speed is independent of onager arm length; it depends only on the angles, spring constant K, and the masses m of the projectile and M of the arm. Equation (1) is an approximation, as usual, and is derived from the assumption that internal ballistics motion conserves energy (from the instant of release right up until the arm thumps into the buffer, perhaps 80 milliseconds later) and that gravitational effects are small. In deriving Eq. (1) I have assumed no pre-stressing of the spring—the thump of the arm against the buffer at the instant of projectile release is quite big enough without the extra force due to pre-stressing.

Engine efficiency is

$$\epsilon = \frac{m}{m + \frac{1}{3}M}\left(1 - \frac{\theta_1^2}{\theta_0^2}\right). \tag{2}$$

Equation (2) immediately tells us that machines that throw heavy stones, with an arm that swings through a large angle, are the most efficient.

Comparing launch speed and efficiency of the onager (Eqs. 1 and 2) with those of the ballista (calculated in the last chapter) for the same parameters, we see that onager projectiles are slower and the machine itself is less efficient than the ballista. Less efficient because its arm is not brought naturally to rest, transferring its energy through the bowstring to the projectile, because the onager has no bowstring—its arm energy is instead transferred to the buffer. The only way that the onager could achieve performance parity with its more sophisticated predecessor is by increasing its spring strength K; this is one reason, in my opinion, why onagers grew from the small machines of Fig. 2 to powerful, heavy and immobile siege engines.[8]

We can confirm that there was an incentive to make onagers big by comparing with ballista dynamics. We construct a mathematical onager with the

[8] More powerful springs meant a bigger kick, which required a sturdier frame, which made for a heavier engine. We have seen that the physical scale (say, doubling the arm length) makes no difference to performance if spring strength is unchanged.

same spring constant as the larger of the two benchmark ballistas of Chap. 6: K = 11,360 Nm. Two of these springs led to a draw weight of 2,500 lb for the ballista; for our onager it is not hard to show that its single spring requires a larger draw weight, 3,000 lb. We choose a release angle of $\theta_1 \approx 40°$ for maximum range (see TA-0: aerodynamic lift, here due to backspin, reduces the velocity angle that yields maximum range). With a light projectile with m = 2 kg the launch speed of this onager is only 84 ms^{-1} (58% of the launch speed for the same missile thrown from the ballista). A 26-kg projectile is thrown at 30.6 ms^{-1}, which is 75% of the ballista speed. The larger benchmark ballista efficiency was ϵ = 82%, you may recall. Onager efficiency depends on projectile mass; from Eq. (2) we find ϵ = 50% for m = 2 kg and ϵ = 87% for m = 26 kg. Most performance parameters are worse for the onager, but become comparable when the two machines—ballista and onager—are big.[9]

Of course the main reason why onager performance is worse is not the relatively crude design, though that is part of it: the main reason is that the onager has only one spring. If we double the initial stored energy of our onager spring to match that of the two ballista springs then we obtain similar ranges. Thus to obtain something like parity with the earlier ballista, fourth-century Roman engineers must have made onagers big and with very powerful springs. Otherwise, they would have been significantly inferior artillery weapons.

The performance of such bucket-onagers could be made better in a couple of other ways [26]. Placing the engine on an uphill slope would mean that, for the same release angle, the arm must swing through a greater arc (i.e., increasing $\theta_0 - \theta_1$), thus increasing launch speed, Eq. (1), and efficiency, Eq. (2). Or the siege engine might be on wheels, with a backstop (but not Marcellinus's stone fortifications) so that it can only roll forward, not backwards. As the arm strikes the buffer, the onager would roll forward a little, slightly increasing projectile launch speed. These are embellishments, however, compared with adding a sling.

3.2 Slings

So they added a sling. Doing so was simple and inexpensive, and led to improved onager performance. A second benefit was that it made a precise and easily adjusted trigger mechanism—in fact, this may have been the main purpose of very short slings, such as those of Fig. 2.

[9] For the calculations of this paragraph I have assumed a throwing arm mass of M = 5 kg, large enough to permit arm dimensions that will survive being thumped by the buffer.

The bad news for us is that the analysis of arm-plus-sling internal ballistics is too complex to be solved analytically, and I can think of no approximations that yield accurate results. I am reduced to setting up the equations of motion of the onager projectile and solving them numerically. Without an algebraic expression for launch speed, we have no at-a-glance insight into the way the dynamics depend on parameters—we just have to do lots and lots of computer simulations and build up a picture. This method of solution is the way forward from now to the end of the book, because all the engines we have yet to investigate have slings—long ones.

For completeness, for the benefit of those readers who are physicists, I include a very brief derivation of the relevant equations in TA-7. These lead to the graphs of Fig. 3 for a small onager of modest power ($K = 2500$ Nm) with arm length $R = 1.2$ m (4 ft) and with a launch velocity angled at 40° to the horizontal, for a high trajectory to project its stone missile over a castle wall. We calculate the launch speed for different stopping angles (shown as 20° in Fig. 3a) and different sling lengths and find the combinations that work best—that throw the projectile at the desired angle with maximum speed.

In Fig. 3a we see how the sling swings as the onager arm rotates: most of the swing is at the end—the sling is accelerating the projectile. This is the benefit of the by-now-familiar double-pendulum effect. There is an optimum choice for arm length and stopping angle which vary with projectile mass. These optimum choices will have had to be found by trial and error back in the fifth century, but once determined it would not be difficult to learn or to make adjustments to the engine, for an experienced crew. It seems likely, therefore, that small onagers will have had adjustable stopping angles, though this may have been difficult to adopt for larger machines. Adjusting sling length would be very simple.

In Fig. 3b,c we see the benefit of the sling for increasing launch speed. The curve for launch speeds without sling were obtained from Eq. (1) for the same onager arm length R and mass M, and the same spring constant K. The optimum stopping angle θ_1 is about 35°. Engine efficiency varies with mass; for the bucket-onager (sling-onager) of Fig. 3b,c it is 18% (32%) for the lightest projectile considered and for both it is over 60% for the heaviest.

This sling-onager can throw a 0.454-kg (1 lb) projectile a respectable 250 m, about 10 m of which comes from the lift generated by projectile backspin. Respectable, and about the same as the range of the smaller benchmark ballista of Chap. 6 despite having a larger draw weight (about 3000 lb). It throws a 2-kg projectile 150 m, which exceeds the range of the small ballista but is much less than that of the larger benchmark ballista. Mixed results—a small onager with

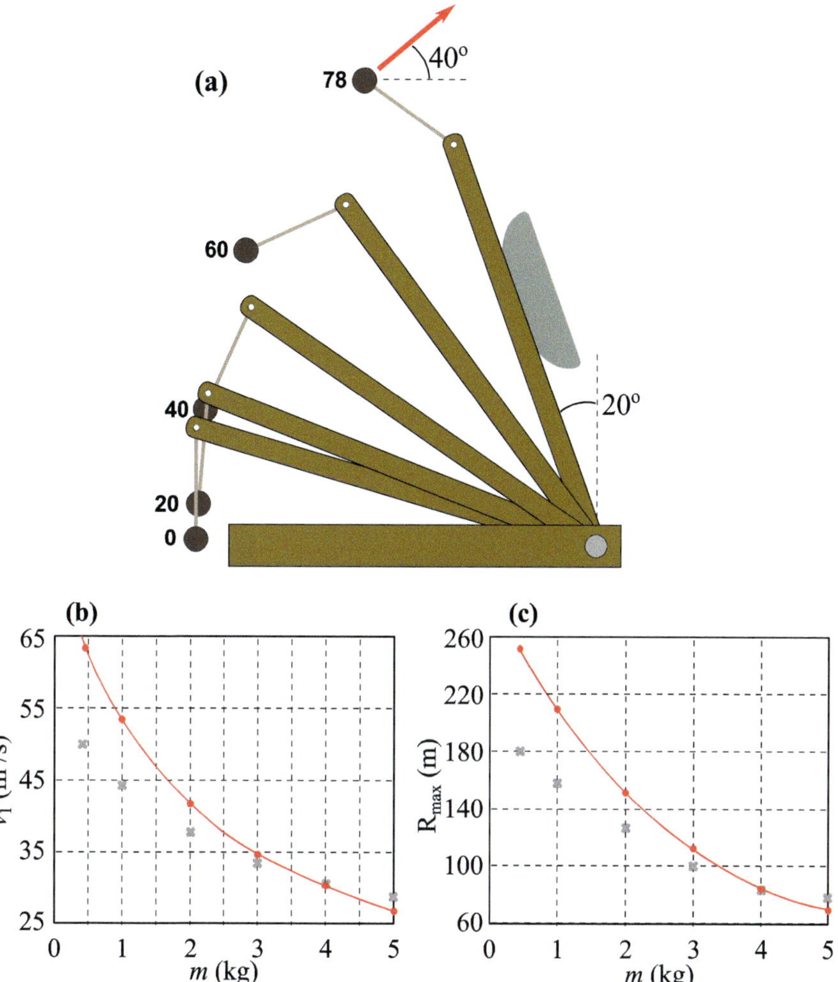

Fig. 3 Onager with sling. **a** Arm and sling position at times (bold numbers) in milliseconds after being triggered. **b** Launch speed versus projectile mass for optimum choices of sling length and stopping angle. Also shown are the results for an engine with the same arm length but no sling (crosses). **c** Maximum range versus projectile mass

a sling works, perhaps well enough, but it is not as effective as a similar-sized ballista.

It is interesting to compare the onager with and without sling for flatter trajectories—say with a launch angle of 20°. This might be appropriate for heavier projectiles intended to damage fortifications; a flatter trajectory will mean a more full-on striking angle rather than a glancing blow. That is, for a vertical wall the onager crew would like to maximize the horizontal component of projectile speed, not the speed itself. (See box ***Trajectories at a glance***). Such a flat trajectory would transfer more momentum to the wall than would a falling trajectory of 45°, say. We find that the optimum sling length for such trajectories is very short, and consequently there is very little difference between the performance of the bucket-onager and the sling-onager. Thus, slings help for long-range shots but not for short-range, flat-trajectory shots.

Given that the sling dynamics are such that short slings are better, and thus that Eq. (1) is a pretty good prediction of the launch speeds that can be generated whether or no the onager has a sling, then we see that the scale of the onager—its arm length, say—is not important. The only reason to increase onager dimensions is to bulk up the throwing arm and frame if the spring has very high power (very large K), so that it does not break when thwacking into the buffer.

To summarize broadly, the only way to reliably increase launch speed and so range and projectile energy/momentum is to increase spring power. Analysis shows that the onager is a step down from the ballista, however you slice it. The Romans must have been hard pressed indeed, to adopt it in place of the ballista.

Trajectories at a glance

In Fig. 4 we see why it might be a good idea, when besieging fortifications, to launch an onager projectile at a low angle (say $\theta_0 = 10°$) instead of at a higher angle (40°). The higher angle gives a longer range for the same launch speed, but it will cause less damage, for two reasons. First, the launch speed is lower, and so the terminal speed, when it strikes the wall, is also lower. Hence less momentum is transferred to the wall, and less damage is done, for the same projectile weight. Second, the low-angle trajectory strikes the wall full on, whereas the high-angle trajectory strikes a glancing blow. However the higher angle may be useful for raining destruction down upon the occupants of the fort.

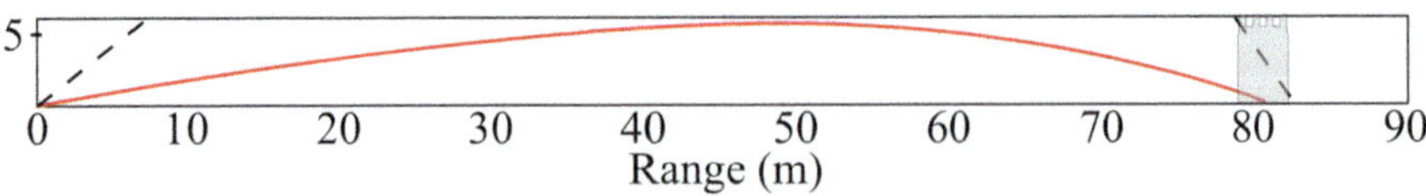

Fig. 4 Target fort at 80 m range: the trajectory with 10° launch angle strikes the wall (gray) nearly full on, whereas the trajectory with 40° launch angle (dashed line) strikes a glancing blow. Same horizontal and vertical scale

4 Traction Trebuchet

Catapults in the West could be considered to be developments of the bow—torsion and then tension machines. In the East, catapults were developed from the sling, powered by human muscles. The ancient Chinese skipped torsion engines entirely, as we have seen, having hit upon a superior technology early on.

The basic idea is illustrated in Fig. 5. A tall frame houses a horizontal axle to which a beam is attached asymmetrically. Historically, the wheels and axle of a cart sometimes formed this fulcrum. From the end of the short side, ropes dangle for the crew to pull on. This action causes the beam to rotate. From the other end—the long side of the beam—hangs a sling with a projectile in its pouch. The action of the pullers causes beam rotation and the sling to swing, releasing the projectile at the desired launch angle.

The idea is simple, and this simplicity is reflected in the design. Traction trebuchets had no complicated parts and were easy to construct, maintain, and use, compared with torsion engines. Their rate of fire was generally higher, and large traction trebuchets were more powerful than any torsion engine. Some skill was required of the crew, who must pull simultaneously upon the ropes to ensure maximum torque is applied to the beam. They pulled downward, being as directly beneath the short end of the beam as geometry allowed. Such a pull direction ensured that most of their power was transferred to the beam, and was not directed sideways, which might destabilize the frame.

To be directly beneath the beam meant that the beam axle had to be quite high off the ground—5 m or more for the larger engines. The higher the axle, the more pullers could be crammed into the space under the beam, and so the more torque could be generated. Each puller could exert a force that was at most his own weight[10] and this limitation provides us with a good indication of the power of a traction trebuchet. Small engines would have a crew of a dozen or fewer; large engines would have a crew exceeding one hundred, as we will see.

4.1 Origins and Spread

Invented in China, most likely in the fifth-century BCE[11] during the Warring States Period, the traction trebuchet was the dominant form of artillery in

[10] It would be interesting to know if traction trebuchet crews attached weights to themselves, so to increase the force they could exert and so increase the power of their siege engine.

[11] There are references from the Spring and Autumn period (eighth-seventh centuries BCE) to a *hui*, which later Han writers interpreted as a catapult.

Fig. 5 Traction trebuchet: a lever on an axle, with a sling at one end and ropes at the other (for the pullers). The pullers were generally underneath the trebuchet, so the axle was high off the ground. Image from the *Wujing Zongyao*, a Chinese military compendium written from around 1040 to 1044. Creative Commons Attribution-Share Alike 4.0 International

East Asia for fifteen hundred years—and in Europe for about 400 years. (It had spread westward across the Old World and reached Europe in the sixth century CE.) This longevity speaks to its qualities: efficient, powerful, and easy to use; it evolved in size, as measured by the number of crew who pulled the ropes. The largest for which we have some written record was an eleventh-century engine in China with 250 pullers (let us call them that). Later, brief and incomplete Arab references tell us of a machine with 400 pullers. Originally more modest anti-personel weapons with a crew numbering in single digits, the later and larger engines were meant for siege warfare, to destroy fortifications. The small trebuchets threw stone projectiles of a few kilograms, whereas their supersized brethren threw boulders of up to 60 kg.

Fig. 6 A modern reconstruction of a small traction trebuchet. Note the slender main beam, with two ropes attached to the short end. Image courtesy of professor Marion Amalric, of Château de Castelnaud, France

Many trebuchets of all descriptions have been reproduced by history enthusiasts and experimental archaeologists alike. In Fig. 6 we see an example. Note that the beam has a structure at the short end (the other end from where the sling is attached) to attach pulling ropes—herein I will refer to this structure as the *rake* since it makes the whole beam resemble one. The few ropes that it dangles from the rake compared with the ancient Chinese engine of Fig. 5 show that it is a small engine. The workings of these two examples are described by the same physical principles. The traction trebuchet is basically a very simple machine with only two moving masses—the beam and projectile, neglecting sling mass—but as always, the devil is in the details, as we will see when we get to the analysis.

The different sizes of traction trebuchets were given different names in China. The "whirlwind" was the lightest, and was placed upon a swivel for easy aiming—clearly this was an anti-personel weapon. The "crouching tiger" was mid-sized, while the "four-footed" was a family of large trebuchets, with trestle frames. Chinese sources state that these machines were easier to transport and set up than the later and larger counterweight machines. Wherever and whenever the source, we are told about the ease of use of traction trebuchets when compared to other types of ancient artillery, and this fact must have contributed significantly to their popularity across the Old World. Thus from

Europeans we are told that the traction trebuchet was more reliable than their torsion engines, because it did not require a torsion spring that was susceptible to damp. Also there was no stored potential energy in the traction trebuchet, and so it was inherently less dangerous to operate. Plus, it worked better: "*The trebuchet was one of the top choices for artillery in ancient and medieval warfare, having the ability to throw heavier projectiles farther than earlier catapults could.*" (Encyclopædia Britannica [27])

It seems that the first encounter between Europeans and the traction trebuchet occurred in 617 or 618 CE during the siege of Thessalonica, by a raiding army of Avars, a nomadic people from the Pontic-Caspian Steppe [28]. Some 50 large engines were used to send heavy stones in high trajectories; descriptions of the engines leave no doubt that they are traction trebuchets. (The history of these machines is blurred by the same uncertainty of nomenclature that we have encountered for torsion engines, but this first European encounter is unambiguous.)

The Arabs adopted traction trebuchets and employed them frequently during the great Moslem expansion of the seventh century CE.[12] These conquests spread the new-fangled siege engines around the European world, as well as establishing Islam outside the Middle East. During the centuries between these initial introductions and the Crusades of the eleventh century, Europeans took up the traction trebuchet and largely forgot the onager. It is a matter of considerable and long-lasting debate among historians as to how long the onager survived—some say it was ditched just as soon as the obviously superior traction engine appeared, while others contend that the onager persisted into the early Middle Ages. However we cut and slice it, the bottom line is that traction trebuchets replaced onagers as the preferred engine for conducting sieges by Europeans, by the eighth or ninth century. They were widely used in medieval Europe and were only partly replaced by their larger, more famous, and more photogenic offspring, the counterweight trebuchet, after the Crusades.

Some accounts of siege engine history assert that traction trebuchets were merely very important weapons across much of the Old World for many centuries, while others claim this to be an understatement—that they also influenced the development of European military architecture[13] and society, and spurred the evolution of science in Europe [1, 29]. Other accounts deny their very existence [20]. Most are in-between [24, 30, 32, 33]—I leave it to the reader to decide the historical place of traction trebuchets in society, and will turn now to their mechanical effectiveness.

---

[12] Some Arab writers referred to large trebuchets with many ropes (see Fig. 5) as *al-Arūs* (the bride), due to the resemblance of the ropes to the long tresses of a woman's hair.

[13] (!) By knocking it down?

4.2 Operational Constraints

The traction trebuchet is a very simple machine (with only two moving masses), and yet there are a surprising number of niceties—subtle or surprising—that complicate the description of its operational effectiveness. The equations that describe the physics are straightforward enough to derive, though complicated, and they can only be solved numerically. I set them up in TA-7 and then run many simulations to reveal the complex interdependencies of parameters, and distill these into a graph or two for you to pour over and ponder.

The surprise (to me, anyway) was just how many independent parameters there are. To describe traction trebuchet motion with reasonable accuracy we need to know all of the following dirty dozen:

- Initial beam angle θ_0,
- Beam length L,
- Beam mass M,
- Distance of the fulcrum from the short end (the rake end) of the beam—expressed as a fraction f of beam length,
- Sling length r,
- Sling initial angle ϕ_0,
- Projectile mass m,
- Projectile launch angle (to the horizontal) ψ_1,
- Axle height H above the ground,
- Number of pullers n,
- Force exerted by each puller F_1,
- Mass distribution of the rake (which affects beam moment of inertia I_b).

I will approach the problem here by building mathematically two engines, one small like that of Fig. 6 and one quite large, like the "bride" of Fig. 5. We will see that they perform differently, and this fact alone explains the diversity of traction trebuchet designs that have been described and drawn in the historical literature. In particular we will see that traction trebuchets—like most projectile weapons—work best with projectiles of a certain weight. That is, the engine is matched to its projectile just as a rifle is matched with the bullets it fires. We will see that the small machine works efficiently when throwing small projectiles and the large machine when throwing large ones, and that these efficiencies fall away either side of an optimum projectile weight.

First I will set up the smaller engine, as it is the easier to do. Most ancient illustrations of traction trebuchets show the engine ready to throw its projectile when the beam is horizontal, and so I choose an initial beam angle of $\theta_0 = 0°$.

This will also be assumed for the larger machine. You might have expected as I did that the best choice for initial angle would be different, with the rake end raised above the fulcrum somewhat, so that the pullers could have a greater angular range over which to exert a torque. Immediately, however, we come across the defining limiting factor of traction trebuchets: human arm length. In practise the pullers can only exert a force on the ropes they pull over a short distance of about a meter, or maybe a little more. Because of this fact, geometry dictates that either the beam is short or the fulcrum location parameter f is small—the beam is highly leveraged, as a fund manager might say—so that the beam can be swung enough to launch the projectile at the right angle.

The beam length is easier to assign. Let us say that our small traction trebuchet has a beam that is 2.85 m long (9 ft 4 in). The sling initially hang limp ($\phi_0 = 0°$, see Fig. 7 for angle definitions). It holds a projectile of mass, well, let us say of any mass up to 10 kg (22 lb). Our small engine is intended to launch its missile to maximum range and so the launch angle is set at $\psi_0 = 35°$. Here we are allowing for aerodynamic drag and lift during the projectile—assumed to be a spherical stone of density 2,500 kg m^{-3}—trajectory to its target. Lift arises here, but not previously for ballista or bucket-onager projectiles, because of the sling: it can impart backspin which generates aerodynamic lift. We will assume that the lift force is a third of the magnitude of the drag force (as for a baseball [31]). Given the spread of projectile speeds that we will find, the aerodynamical equations of TA-0 tell us that the launch angle for maximum range is within a few degrees of our chosen value.

The axle height is set at $H = 3$ m above the ground. This is sufficient so that the swinging beam will not decapitate the pullers as it passes over them (recall that they are beneath the rake end). Our small engine is powered by $n = 8$ pullers. The force they exert cannot exceed their own weight, and only then if they are located such that the rope they pull is perpendicular to the beam, for maximum torque. The average Chinese man today (I imagine that my mathematical trebuchet is powered by mathematical Chinese pullers) weighs 70 kg; his Han antecedents will have weighed less. The average pull they exert will be a little less than this, so I have settled on a force of $F_1 = 500$ N. The real average force exerted by real pullers of real traction trebuchets will not have differed very much from this value.

Mine is a light machine with a strong but slender beam like that of Fig. 6. I will assume it needs no reinforcement and that the rake structure adds little to the beam mass. The beam is taken to be a circular tree trunk of constant width: given the density of wood and the bending forces that apply, I estimate that a beam with diameter 6 cm ($2\frac{1}{2}$ in) and mass $M = 5.5$ kg will do the job. Two parameters remain to be fixed. Repeated simulations lead to the conclusion

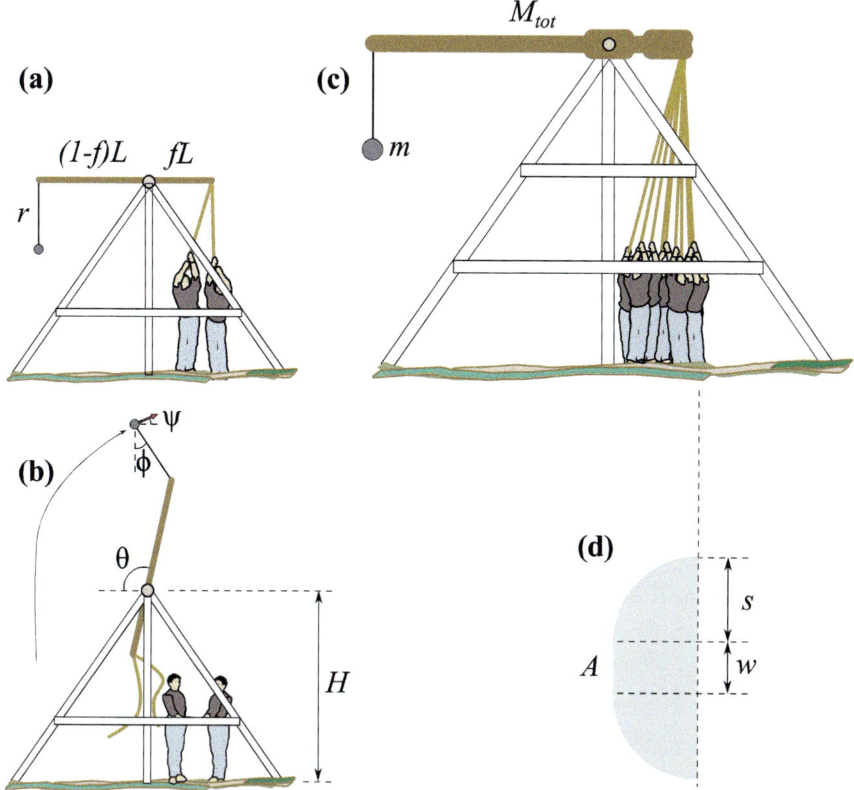

Fig. 7 Traction trebuchet models. **a** A small engine, drawn to scale, before the ropes are pulled... **b** ...and just as the projectile is launched. **c** A larger engine, drawn to the same scale. **d** Viewed from above, the area beneath the rake end of the beam on which pullers can stand. The size of this area determines the number of pullers and so the force, or "draw weight" of the engine

that the small traction trebuchet performs optimally with a sling length of $r = 0.35L$ and with a fulcrum that is a distance fL from the rake end, where $f = 0.35$. (Thus $fL = 1$ m, so that the pull distance—the draw length, in crossbow parlance—is $d = 1$ m. Thus the pullers can torque the beam until it is vertical.) These values are close to optimum for a spread of projectile masses up to 10 kg, and is the best choice that can be made unless the machine is designed for a very specific value of m, in which case both r, f could be tweaked a little to add a few percent to missile launch speed and range.

Such is our small traction trebuchet—laid out approximately to scale in Fig. 7a,b. We will see how it performs in the next section. In the rest of this section I will describe my larger mathematical engine. This engine is defined in similar manner, with a slight complication. In short, we specify draw length

$d = fL = 1.2$ m, $f = 0.24$ and so $L = 5$ m, $r = 1.5$ m, $M = 44$ kg and $n = 68$. Projectile masses are spread over the range $m = 10 - 50$ kg. The axle (the beam fulcrum) is $H = 5$ m off the ground, and here is our slight complication.

From Fig. 7c you can see that the crew members are crammed together beneath the rake—they cannot be spread out wide because the angle of some of the ropes would then become ineffective at applying torque to the beam. If we require the ropes to be within, say, 30° of the vertical, then the area on the ground where the pullers can stand is as shown in Fig. 7d.[14] Because the beam swings to the vertical during a launch (as suggested in Fig. 7b) the ropes are confined to an area beneath the initial horizontal beam, not out in front of it, which would lead to ropes pulling in the wrong direction near the end of the beam trajectory. The area shown depends upon the rake width w and the radius $s = H \tan \beta$ where β is the maximum permitted rope angle (we have fixed this at 30°). Thus the area is calculable. If the density of pullers that can be crammed into this space is $\sigma = 4$ per square meter then we can calculate the number of pullers, and so the power of the trebuchet. For our large trebuchet we have a draw length of $d = fL = 1.2$ m (the maximum feasible by average-sized humans) and $w = \frac{1}{4}L$ which gives us $n = 68$. Repeated simulation shows that the optimum performance attains for $f \approx 0.24$ and so $L = 5$ m. The mass of such a beam is $M = 44$ kg. Again simulation provides the best choice for sling length r given these other parameter values: we find $r = 1.5$ m. Thus we have our larger traction trebuchet.

In analyzing this engine it dawned on me that the procedure for specifying engine dimensions follows a Philon-like procedure based upon the power source. Here axle height determines n, and for the best choice of f we derive the remaining lengths.

It would be easy for querulous readers to quibble with my parameter choices ("That beam mass is too small," "the beam mass is too big," "that beam is flexible, not stiff," "the sling is elastic, not stiff," "you have assumed a massless sling—wrong!") but only one of these troubles me. It is known that the slender beam was elastic enough to be exploited. Thus there are historical images which show a traction trebuchet crew member hanging on to the end of the sling during launch—he would let go very shortly afterward [1, 32]. This action would cause a flexible beam to bend and rebound, so boosting the projectile speed. My analysis does not include this effect. Note that its efficacy relies upon a coincidence of parameters: the elasticity of the beam needs to be such that it rebounds just at the point of release of the projectile; the

[14] This derivation of the number of pullers by calculating the area to which they are restricted by geometry was first presented by Chevedden [1], and here adapted.

time of this release depends on projectile mass, I find, and so this trick will work only for certain projectile masses (and would actually be a hindrance for other masses). Presumably the crews of historical traction trebuchets found out which projectiles benefited from a dangling crew member by trial and error, not having the benefit of a computer simulation to aid them in the decision.

Now it is time to see how these engines performed. I end this section with a note to readers to bear in mind that my larger engine is certainly not the largest possible—so what is the largest? We will answer this interesting question later.

4.3 Performance

I have taken to call my smaller traction trebuchet "the 35" in the same way, I suppose, as firearms are sometimes given identifying numbers (as in "the French 75" or "the AK47") though for a different reason—nothing to do with caliber or year of introduction. (The smaller trebuchet has an axle position such that $f = 0.35$, a sling length such that $r/L = 0.35$ and an optimum launch angle of $\psi_1 = 35°$.) It throws a 1-lb projectile 380 m, or nearly a quarter mile, and a 10-kg projectile about 70 m. The ranges achieved for masses between these limits are shown in Fig. 8. These results compare favorably with the torsion engines, adding to the reasons why traction trebuchets were considered to be superior siege engines, replacing the torsion powered machines in Europe when they arrived from the East.

We define the efficiency of a traction trebuchet to be the fraction of energy provided by the pullers[15] that is taken away by the projectile. There are two noteworthy features concerning efficiency here: it is high for both machines, and it peaks for a specific projectile weight. Thus, the smaller trebuchet is most efficient when it throws missiles of two or three kilograms, and the large one when throwing missiles of 10–20 kg. There are suggestions here of a feature found earlier for other projectile weapons: the projectile is matched to the weapon. Here the spread is quite broad, just as a slinger can throw a wide spread of sling bullet weights.

An upper bound is shown in Fig. 8a for launch speed, that provides us with an algebraic expression approximating traction trebuchet launch speed. Thus, if we assume perfect efficiency then the initial kinetic energy of the projectile equals the input energy, from which we obtain for launch speed the expression $v_1 \le \sqrt{2nF_1d/m}$. Because efficiencies are high, this upper bound is in fact a reasonable estimate, as we see in the figure.

[15] This energy is approximately equal to the number of pullers multiplied by the force exerted by each puller multiplied by the "draw length": $E_0 \approx nF_1d$.

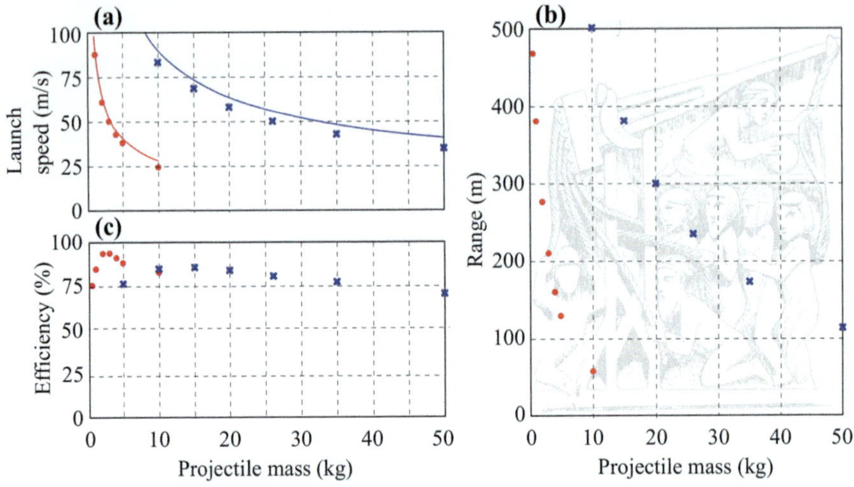

Fig. 8 Trebuchet performance data. **a** Small trebuchet ("the 35") launch speed versus projectile mass (circles). Crosses are for the larger trebuchet. Solid lines: upper limits assuming perfect efficiency. **b** Range versus projectile mass. **c** Engine efficiency versus projectile mass

The 68 pullers of my larger engine can send a 1-talent stone projectile 230 m, as we see in Fig. 8b. This beats Philon and his ballista. What about bigger traction trebuchets? What about REALLY big ones? There are Chinese and Arab written records about such machines, varying considerably in detail, but with enough information for us to make a stab at their capabilities, though not enough information—that I am aware of[16]—to reliably construct a mathematical model such as that for my "35." So here is a plausible estimate for the 250-man traction trebuchet mentioned in a Song Dynasty (eleventh century) military compendium called the *Wujing Zongyao*. Dimensions are given, and from calculations similar to those of Fig. 7d it has been shown that they were of such a size that $n = 250$ is just about possible. From Fig. 8 we might estimate the efficiency of this giant to be perhaps 85% and so from our handy formula given in the last paragraph we come up with a launch speed of about 112 ms^{-1} for a 1-talent spherical stone projectile. Calculations like those of TA-0 tell us that this projectile, launched at an angle of $\psi_1 = 35°$ will be thrown 830 m, or half a mile. Repeat: HALF A MILE. Interestingly, the same formula and calculations suggest a range of 436 m for a 60-kg projectile, which is, ahem, consistent with the "at least 75 m" range claimed in the *Wujing Zongyao* [1]. At this point I must remind readers that the calculation assumes perfect effi-

16 If a reader has credible information about the dimensions of these supersized traction trebuchets, please let me know and I will build one—mathematically.

ciency, and we can infer from Fig. 8c that efficiency decreases with increasing size. Thus, take my predictions for the 250-man trebuchet with 1 talent of salt.

I have really pushed the envelope with this 250-man engine by assuming that each puller can exert his full weight as his pulling force, due to them being packed like sardines beneath the trebuchet fulcrum. They are so well trained that they can pull simultaneously on their ropes, and dangle like a bunch of grapes while the beam and sling do their thing.

Even larger machines are mentioned by twelfth-century Arab sources, with even fewer technical details provided. I will end my discussion of traction trebuchet giants here because, even if they existed in the early Middle Ages, and even if their engineers pushed the envelope as I have suggested for the Chinese machine, their performance was soon to be surpassed by the superstar of ancient projectile weapons—the counterweight trebuchet, of course—which will be introduced in the next section.

There was an interesting variation of the traction trebuchet which might be regarded as a precursor of the counterweight machine or as a hybrid between them. Some large traction machines had the rake end of the beam deliberately weighted so that the beam was balanced without a load. This feature would have made the machine easier to handle—for example when loading the projectile into the sling pouch. On the other hand it will have led to a less efficient machine, because the increased mass will have diverted puller energy from the projectile.

5 Counterweight Trebuchets

The counterweight trebuchet was very like its older traction relative except at the short end of the beam, where a large weight replaced the ropes and pullers. This weight was either fixed directly to the beam end or else hung off it, from a hinge, like a pendulum. NB: the natural, though very clumsy, names for these trebuchets would be "fixed-counterweight trebuchet" and "swinging-counterweight trebuchet." This nomenclature is too clumsy even for me and so I will herein denote them "FCWT" and "SCWT" respectively (see Fig. 9). Unsexy but concise.

These engines were the heavy artillery of Old World armies for perhaps four centuries and yet, surprisingly, we don't know where or exactly when they were invented. Best guess is the Eastern Mediterranean at the end of the eleventh century, probably by Byzantines or Franks ("Latins," as the Byzantines called them). Certainly it was in this region during the First Crusade that we have our first accounts of the new type of engine. Judging by the rate at which it spread

Fig. 9 Two types of CW trebuchet. In the foreground is an SCWT; immediately behind it is an FCWT. They both look to be of medium size, with counterweights of a few tons. Thanks again to Marion Amalric of Castelnaud Chateau, France, for permission to reproduce this image

during the twelfth century, and the extent, it must have been an immediate hit (no pun intended). We will get to the history shortly, but first I would like to opine about its intellectual origin: *how* did its unknown inventor come up with the idea of a counterweight trebuchet?

5.1 How and Why

The traction trebuchet had been the main siege engine in the East since its invention by the Zhou Dynasty Chinese in the fifth-century BCE. It spread westward and replaced the onager, as we have seen, and so was used across the Old World by the end of the first millennium, so most certainly will have been known to the military engineer (I presume) who would go on to invent the counterweight (CW) trebuchet. Another machine that will have been known to him[17] is the *shaduf* (one is shown in Fig. 10). This venerable crane could be seen on many a riverside across much of the Middle and Far East, and in North Africa during the period of the First Crusade and for more than a thousand years beforehand. It was used to lift water from the river to the shore. In crane-speak we have a tower upon which balances a horizontal jib, from the long arm of which a hook is suspended, and from the short arm a counterweight. The hook holds a bag which is lowered into the river to fill with water. Then the jib is lifted and swung around so that the water bag can be lowered to the ground.

The inventor of that first CW trebuchet must have been thinking about how to improve upon the traction trebuchet. First and foremost its main limitation is size and the maximum weight of projectile that it could throw. So, it seems to me, our military engineer was sitting on a riverbank one day while his assistants were training a new bunch of woodenheaded recruits in the fine art of trebuchet rope-pulling, wondering if there might be a better way of throwing rocks. He thought about the hybrid traction trebuchet, and his eye fell upon a *shaduf* being operated nearby and...and...nothing. He put two and two together and got two. Clearly the juxtaposition of traction trebuchet and *shaduf* was insufficient to inspire the CW trebuchet. We can be sure of this because the two had been occupying the same regions, the same large tracts of Eurasia (the Middle East, Persia, central Asia, China) for a thousand years and yet none of the people in the armies of those vast regions had thought of it, despite a universal perception that they needed to throw bigger rocks.[18] The Chinese had invented the traction trebuchet way back when, and at the time that the CW trebuchet was born the Song Dynasty was ruling China—a dynasty known for its military inventiveness (think gunpowder weapons)—and yet, even for the Song, that juxtaposition was insufficient to swing it (no pun intended). The "bride" was not about to marry a mere water carrier.

The shaduf may have inspired the traction trebuchet much earlier in history, but the two of them were not enough to inspire the CW trebuchet. Enter the onager—a crude beast and an unruly brute. It and the traction trebuchet coexisted in the Balkans and Anatolia and the Middle East at the time that the CW trebuchet was conceived. Yes, I think that there was an unholy union between the "bride" and the beast, the bastard child of which was the counterweight

[17] Almost certainly the right pronoun, though there is one female who appears very early in the written accounts. Anna Komnena was a Byzantine princess who wrote about the reign of her father, Alexius I. She mentioned siege engines and trebuchets, and some scholars believe that the CW trebuchet developed in Byzantium during his time. Or maybe he was on the receiving end, during his wars with the Seljuk Turks in Anatolia. Or he learned of them from his interaction with the Arab people. Or...

[18] I am being flippant here, of course. I mean that there is and always has been among every country of the world at all times in history a desire and motivation to make bigger weapons.

Fig. 10 Shaduf. Adapted from a Public Domain image found at DT Online

trebuchet. Our engineer on the riverbank, annoyed at his team of recruits, may have been looking at a shaduf or a traction trebuchet swinging its heavy load, or a hybrid trebuchet, or maybe he was actively involved in a siege at the time he had his Aha! moment, directing both traction trebuchets and onagers against a castle wall that was too strong for them both. He wanted to eliminate the dreadful recoil of the onager; he wanted to increase the weight of projectile that his traction trebuchets could throw; he saw a shaduf lifting a heavy bag of water using its counterweight and...and...two and two made four.

5.2 Comparison of Trebuchet Types

If it was a large FCWT that threw the projectile of Fig. 11, then it could have flown some 230 m (further than this, had the catapult been a SCWT—of which more later). A large traction trebuchet could not have managed half that distance, and even the largest onager or Archimedes' ballista could barely have thrown it off the cliffs at Syracuse. Yet as we will see the traction trebuchet was used alongside the CW trebuchet for centuries and was never completely replaced by it. So how did the two catapult cousins compare? And why were there fixed-CW and swinging-CW versions of the CW trebuchet? In fact there were more than two versions, as we will see in the paragraphs to follow.

Fig. 11 Stone catapult projectile ($m = 66$ kg, $14\frac{1}{2}$-inch diameter) found near Montfort Castle, N. Israel. Probably used by Mamluk forces besieging the castle in 1271. Image courtesy of the Metropolitan Museum of Art Open Access policy

We saw in the last section that the power stroke or draw length of the traction trebuchet is limited by the length of the human arm to a little over a meter, at most. Consequently the power transferred to the beam is limited, and so the energy of the projectile at launch is limited. There is no such restriction for CW trebuchets—the power stroke is much longer. Thus CW trebuchets are superior for long ranges (Dynamo CW 1: Traction United 0) and can throw stone balls that would be too heavy for a traction engine (2:0). On the other hand, the fact that a heavy CW is fixed to the beam means that it moves horizontally (front-back) as well as vertically during the beam rotation. So what? The vertical component transfers gravitational potential energy from the CW to the beam and so to the projectile, but the horizontal, sideways movement does not—it just diverts energy from the projectile. Thus we can expect CW trebuchets to be inherently less efficient than traction trebuchets—traction engines are back in the game at 2:1. We will see shortly that this wasted energy is the main difference between FCWTs and SCWTs; for the latter the sideways movement of the CW is less and so there is less wastage of stored gravitational energy. For two identical engines differing only in the manner of CW attachment, the swinging engine will throw a projectile further.

Traction trebuchets are lighter in weight; the CW machines need sturdier frames to withstand the forces that arise due to CW movement during the launch phase. This fact means that CW trebuchets are more expensive and time-consuming to construct, and more difficult to transport to the siege location. (Dynamo 2: United 2). The large CW takes effort and time to raise to its initial (trigger) position. Thus the traction trebuchet could throw more projectiles per hour than its CW cousin—it had a superior rate of fire. Traction United takes the lead, 2:3.

One of the reasons why trebuchets took over from torsion engines was, recall, the smoothness of the throwing action—especially when compared with the onager. This smoothness gave trebuchets a tactical advantage, as well as a longer lifetime and fewer maintenance issues. In the context of a siege, once the crew of a trebuchet had found their target—say a rampart or a battlement they wished to destroy—then they could reliably hit it again and again. This feat would be more difficult to replicate for a ballista and impossible for an onager, which needed to be reset after every shot. Between the three types of trebuchet, both of the CW versions would be better able to repeat a shot more easily than could a traction trebuchet, due to unavoidable variations of the pullers' actions from shot to shot. And the score is tied, at 3:3. Between the two types of CW engine that we have considered, the SCWT might perhaps be marginally less able to repeat a shot exactly than a FCWT, because it has more moving parts and so more sources of natural variation.

If there is a tie-breaker, then perhaps it comes from a consideration of the crews. Their longer ranges might suggest that CW trebuchet crews would come under less intense fire from enemy arrows than would traction crews. The effort involved in loading would be greater for CW crews (raising the CW to the trigger position) but traction crews may tire more over time due to the greater number of missiles thrown. The training of a traction crew might be more demanding than that of a CW crew.

However we cut and dice these considerations, it is clear to me that the performance difference between these two types of trebuchet—the difference between their efficacy as siege weapons—is not very great. I can see why both persisted. If very heavy rocks were called for, then only CW engines could do the job; if a high rate of fire to suppress enemy soldiers at some key position was required, then traction engines would be used. If scaring the bejesus out of the citizenry of a besieged city was the aim, then a 60-kg stone ball dropping through the roof might be more effective than six 10-kg balls.

5.3 CW Evolution and Spread

Wherever, whenever and however it originated (my best guesses: late tenth or early eleventh century, southeastern Europe or Anatolia, siege engineer with knowledge of onager as well as traction trebuchet) the CW trebuchet expanded the success of earlier horizontal-axle siege engines. Expanded in two senses: CW trebuchets were more powerful than traction engines and so could send spherical stone projectiles of a given weight to greater ranges, and CW trebuchets could throw heavier projectiles that could damage battlements (maybe even curtain walls, though this is disputed) that were impervious to traction trebuchet projectiles.

The earliest written records, and illustrations, of CW trebuchets are from the Near East and environs during the early Crusades. A twelfth-century Arab text (by Mardi ibn Ali al-Tarsūsī, writing a military manual for sultan Saladin) refers to an engine that was clearly quite small, requiring only a single crewman, and serving the double purpose of catapult and siege crossbow winch (that is, the CW trebuchet was used to draw the bowstring of a large crossbow). Fast forward 220 years and we read the account of a German military engineer (Conrad Kyeser, in his book *Bellifortis*, 1405) who tells us of an enormous machine with a beam that is over 15 meters long (50 ft), and with an axle that is 11.5 meters (38 ft) above the ground. Chevedden [32] writes of such enormous machines with counterweight boxes the size of a peasant's hut and containing weights of "tens of tons."

Clearly there was an evolution going on here. A good idea emerged and, spurred on by the necessity of war, was developed in different ways relatively quickly. By the time of Kyeser's writing, use of the CW trebuchet had spread across the Old World both east and west—as far east as China and as far west as Britain and the Atlantic coast. The Chinese referred to it as *Huihui Pao* or "Muslim trebuchet"; it reached China during the Song Period via the Mongols, who themselves acquired it from Muslim siege engineers further west. It was used alongside the traditional traction trebuchet. King Edward I of England constructed "Warwolf," a very large CW trebuchet, in 1304 to besiege Stirling Castle. Supposedly the Scots surrendered upon setting eyes upon this monster being assembled, but Edward wanted to see it in action, and so he would not accept their surrender until it had sent a single stone through two castle walls "like an arrow through cloth."[19] This account sounds contrived to me, though the machine did exist.

There were many steps on the way, in the design evolution of CW trebuchets as well as the far and wide spread of their use. Initially the CW trebuchet was small and resembled a shaduf in that the counterweight was fixed to the short end of the beam. Later the swinging counterweight was used due to its somewhat better performance, which we will investigate in the next section. A small variant was the *bricole* or *couillard*, placed on a pole with a split CW, the two parts of which swung past the pole (see Fig. 12). But this type of engine

[19] Wikipedia article *Warwolf* accessed July 5, 2025.

Fig. 12 In the foreground is a *couillard*. The name is the French word for an uncastrated animal—look at the twin counterweights. This small trebuchet could launch projectiles of up to 50 kg. In the background is a larger fixed-CW trebuchet. Image courtesy of Marion Amalric of Castelnaud Chateau

was a sideline: the main reason for the existence of CW trebuchets was to do what traction trebuchets could not: throw very heavy stones at fortifications. To this end CW trebuchets were usually built on site rather than transported, due to the size and weight of their frames, and were aimed at a specific target, where the repeatability of its shot gave it an advantage in bringing down walls. The direction of aim could not be changed easily or quickly (except perhaps for a small couillard).

It is a matter of some debate in the academic literature about the effectiveness of CW trebuchets throwing rocks against castle walls. We will see that the largest of these beasts could hurl a 300 kg rock several hundred meters, which means it could certainly have damaged walls and, how to put it, gained the full attention of the besieged. I have alluded earlier to an arms race that took place between siege engines and castles, each growing as a result of the other's growing. On the siege engine side, traction trebuchets became larger, and then FCWTs became larger than those, and then SCWTs became giants, over centuries. On the castle side, these fortifications—in Europe—consisted of rammed earth and timber before the ninth century, becoming stone in the tenth and eleventh, growing larger with towers through the thirteenth and spreading with concentric rings of walls and towers into the fourteenth. There is no doubt of a correlation between the growth of castle size and siege engine size.

The simplest view is that, as the weight of stone shot thrown from trebuchets increased, so the strength of castle walls increased. The same argument could be extended to say that the addition of outer concentric walls was to counter the increased range of trebuchets. This view is probably too simplistic, however. Fulton points out that the details are wrong. Thus, if large CW trebuchets could knock down the main curtain walls of castles, then the response by castle builders should have been to increase wall thickness. This increase did indeed occur, but mainly after the introduction of gunpowder siege weapons. These bombards and early cannon fired missiles with an order of magnitude higher launch speeds and therefore an order of magnitude higher momentum and two orders of magnitude higher energy. Their missiles flew in a flatter trajectory and so would strike the walls full on, or more nearly so (i.e., with the missile velocity direction more-or-less perpendicular to the wall). But the growth of castles began with an increase in the height of walls, and not a major increase in thickness. The implication is that even the heaviest CW trebuchet would have difficulty knocking down curtain walls, even with multiple shots at the same point.

A more nuanced interpretation of the correlation between trebuchet and castle growth is perhaps justified. The larger trebuchets could knock down more walls than could smaller trebuchets—towers, battlements, ramparts, but not the main curtain walls of a large castle. These were breached by other means, when the siege was successful (that mighty tool from the dawn of warfare—the siege ladder—being principal among them[20]). We have already seen that the trajectory of siege engine projectiles is not flat, and indeed at longer ranges it will arrive at ground level moving more nearly vertical than horizontal. Thus, it could be argued, large trebuchet projectiles would do more damage to horizontal surfaces (such as the walkways atop battlements, or roofs inside the walls) than to thick vertical walls. It would be interesting to know if any of the primary sources mention trebuchets being located below a castle, so its projectile could strike the castle wall at the apex of its trajectory when it is moving horizontally.

The later, largest engines had several bells and whistles added, like luxury packages added to new cars. Thus CW1.0 might denote your basic, entry-level FCWT, evolving up to, say, CW1.3 over the years as size increased. Meanwhile for the more discerning poliorcetes[21] you might consider CW2.0 s, with its

[20] Along with mining, guile and treachery—and intestinal fortitude within the leading besiegers.

[21] Besieger, or besieged, or those who study sieges.

innovative swinging counterweight, or even (if your purse will stretch) the latest CW2.1sw, bigger and set on wheels as well as with a new and very cool swinging counterweight. Would your majesty like to take it for a spin?

It is possible that the elites of the early Middle Ages regarded such high-profile weapons as something of a display (Edward I of England comes to mind) permitting them to showboat their power and impress their fellow rulers. In a later, more refined age this kind of thing happened with mechanical watches: the basic machine was improved somewhat by very complex and expensive additions that were more an exhibition of wealth and style than a need for accurate timekeeping. (In the modern age, flashy cars are the best analogy I can think of.)

It is interesting to note that these bells and whistles did improve trebuchet performance—especially the swinging counterweight. Now it is time to look at that performance to see why CW trebuchets were so much more powerful than anything that had gone before, and to see why the add-ons helped, to a greater or lesser extent.

5.4 Performance

The equations of motion for a fixed-CW trebuchet are given in TA-7 along with an outline of their derivation. Presenting you with their solutions involves the same kind of enjoyable (for me) complications that arose for traction trebuchets: juggling lots of parameters. So how best to convey the overall picture, the gist, the quintessence of CW engines? Consider Fig. 13.

In Fig. 13a we have an FCWT with the sling initially dangling vertically. When the trigger is released the action is as shown, until the projectile is released at an angle of 45° or so, for maximum range. If the sling is long, then either the axle must be high off the ground, or else there must be a trench dug so that the swinging projectiles do not drag along the ground. In Fig. 13b we have an SCWT with smooth rails for the projectile to slide along prior to lift-off. The difference in performance between these two variants is mainly due to the CW motion. The fixed counterweight rotates in a circle, whereas the hinged CW swings freely and, with other parameters well chosen, falls more-or-less vertically, as here. Thus when the projectile is released it takes away more of the initial stored energy with it than does the FCWT projectile, because at the instant of release the swinging CW has less speed and therefore less energy. We will quantify this statement shortly, by comparing launch speeds for two CW trebuchets. For now note that, post release, the swinging-CW swings back in such a way as to reduce beam rotation rate and thus bring the engine to a gentler stop than occurs for the FCWT.

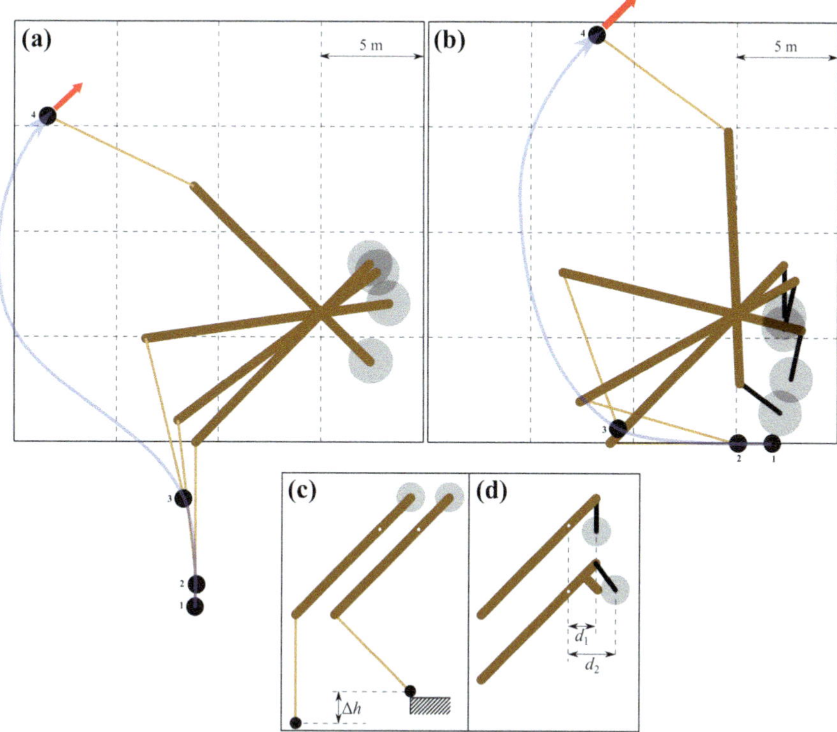

Fig. 13 Counterweight trebuchet variants. **a** Fixed CW, with hanging projectile. **b** Sliding CW, with sliding projectile. **c** Fixed CW, showing propped projectile. **d** Swinging CW, showing propped CW

The two engines of Fig. 13a,b were chosen to have the same beam length, sling length, and the same projectile mass and release angle (velocity direction indicated by the arrows). The resultant trajectories are shown at the beginning of each third of the trajectory duration t_1. Thus, positions labeled "1" are at the start, those labeled "2" are at $\frac{1}{3}t_1$, those labeled "3" are at $\frac{2}{3}t_1$, and positions "4" are at release. Note the very different internal (i.e., before release) trajectory of the projectile, and of the beams.

In Fig. 13c we see an FCWT with a dangling projectile and one with a "propped" projectile. The latter has more initial gravitational potential energy than the former and so, optimizing other parameters, we can expect it to perform better. That is, we can expect that it will release its projectile at the desired angle with higher launch speed—and this is the case. Similarly in Fig. 13d we see a SCWT in its initial position with a dangling CW and one with a "propped" CW. The latter exerts more torque about the axle (by a factor d_2/d_1) than the former and so we can expect it to perform better. It does.

This variant was known to medieval siege engineers. Propped counterweights are a little tricky because the other parameters have to be chosen just right to squeeze out the higher projectile launch speed—choose them wrong and the performance can be worse.

This introduction has been aimed at conveying the variations in possible configurations of the two main types of CW trebuchet. Of course there are also all the parameters we found for traction trebuchets to consider: for a given beam length, counterweight mass and projectile mass, how should we choose axle position f and sling length r? By trial and error, a medieval engineer would say. The modern equivalent is repeated computer simulations. We see the results for a small and a large FCWT in Fig. 14. The small engine parameters are: counterweight mass $M_{cw} = 5000$ kg, beam length $L = 5$ m and mass $M = 92$ kg and initial beam angle 45°. For different projectile masses, I vary the sling length and axle position to find the optimum combination (see how hard I work for you?). The big machine parameters are $M_{cw} = 10000$ kg, $L = 12$ m and mass $M = 630$ kg. (The beam is much sturdier due to the higher torques, of course.) The release angle is the same.

Results? First we note the shock and awe of a Greek ballista engineer or a Roman onager crewman at the capabilities. Launch speed of 46.4 ms^{-1} for a 100-kg stone? That's a range of over 200 m. Yikes. We know that the improved performance is due to the much larger stored energy—gravitational potential energy rather than the elastic potential energy of sinew. The 68 traction trebuchet pullers for our larger engine would have been impressed. We know that CW rotational energy imparted to the beam (and so to the sling and so to the projectile) exceeds that of a traction trebuchet due to the limited traction power stroke.

In Fig. 14c, d we see the necessity for optimization. For fixed r there is an optimum f and vice versa. (In these simulations I have assumed a projectile prop angle of 45°.)

Having confirmed the superior performance of CW trebuchets against other siege engines, and conveyed to you something of the interplay between parameters that goes on during the dynamical launch phase, I should pause so that we can consider the medieval engineers and crew who designed and operated these machines—given the number of parameters they must have invested huge effort to find the best combinations. This accumulated knowledge must have elevated the cognoscenti in the eyes of their generals and kings. And now we turn to the main trebuchet performance comparison: between the fixed-CW and swinging-CW variants.

The comparison is not quite straightforward, but we can obtain meaningful results by proceeding as follows. Let's fix $M_{cw} = 4000$ kg, say, and $L = 10$

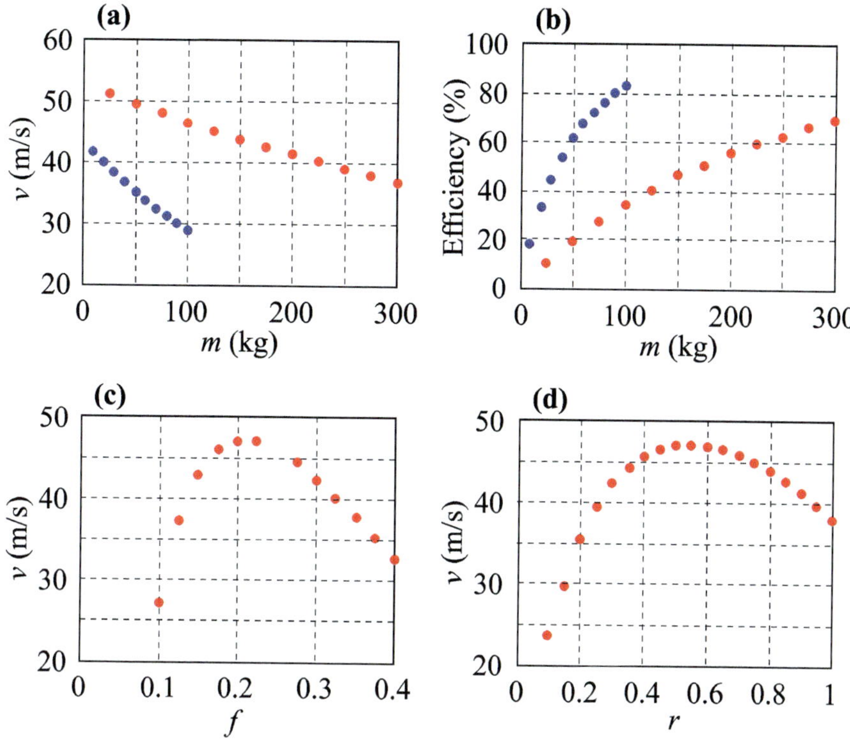

Fig. 14 FCWT performance, for a small and a large engine. **a** Launch speed versus projectile mass. The maximum projectile mass for the small engine is 100 kg. **b** Machine efficiency versus projectile mass. **c** For the big machine with $m = 100$ kg and $r/L = 0.55$, launch speed versus axle location f. **d** For $f = 0.2$, launch speed versus r/L

m. Let us throw two weights of projectile, one with mass $m = 70$ kg and another with $m = 30$ kg. Then we optimize the remaining parameters as just described, and see what happens. It turns out that for the FCWT the largest launch speed that can be achieved, assuming a launch angle of just under 45° for maximum range, is $v_1 = 49.7$ ms^{-1} for the 70-kg projectile. This best value arises when $(f, r/L) = (0.28, 1.10)$. For the SCWT we find $v_1 = 51.7$ ms^{-1} for $(f, r/L) = (0.29, 0.64)$. As we anticipated, the SCWT performs somewhat better. As for the lighter projectile we find (fixed) $v_1 = 58.5$ ms^{-1} for (0.24,1.10), and (swinging) $v_1 = 62.5$ ms^{-1} for (0.27,0.66). In terms of range, the SCWT sends the heavy missile 250 m (233 m for FCWT) and the light one 345 m (303 m for FCWT). To ensure that like is being compared with like as much as possible, for this head-to-head I have assumed a FCWT projectile prop angle of 90°, so that the sling is initially horizontal for both CW trebuchet types.

As for other variants, I will leave these to the interested reader—this chapter is quite long enough already.

6 Summary

Siege engines started out small: slings and crossbows—then they started to grow and eventually became huge, along with the fortifications they besieged. In Europe crossbows led to torsion engines and in East Asia slings led to traction trebuchets. The superiority of the latter became evident when they spread westward and took over from torsion engines in Europe. For reasons other than technical superiority, the onager had replaced the ballista in Europe. Where all these engines overlapped at the turn of the first millennium—onagers and traction trebuchets coexisted in Eastern Europe and Anatolia at this time—a variant trebuchet was created. The CW trebuchet grew very large over the next four centuries, until it was superseded by bombards and other gunpowder siege engines.

Counterweight trebuchet design flowered, in that several variants arose. The growth in size of these machines and that of castles may have been a kind of arms race, a symbiosis that improved both, driven by the other. All the technical developments were empirical, as no scientific analysis was yet available, though some historians regard trebuchets as spurring the development of scientific methods. More prosaically, the effects of the largest bad-ass SCWT were likely psychological as well as physical, permitting kings and generals to peacock while inducing terror in those beneath the falling projectiles.

References

1. P.E. Chevedden et al., The traction trebuchet: a triumph of four civilizations. Viator **31**, 433–486 (2000). https://doi.org/10.1484/J.VIATOR.2.300774
2. K.D. White, *Greek and Roman Technology* (Thames and Hudson, London, 1984) p. 59
3. A. Caliebe et al., Insights into early pig domestication provided by ancient DNA analysis. Sci. Rep. **7**, 44550 (2017). https://doi.org/10.1038/srep44550
4. A.E. Dien, The stirrup and its effect on Chinese military history. Ars Orientalis **16**, 33–56 (1986)
5. M. Denny, *Float Your Boat: the Evolution and Science of Sailing* (Johns Hopkins University Press, Baltimore, MD, 2009)
6. M. Korfmann, The sling as a weapon. Sci. Am. **229**, 34–42 (1973)
7. C. Harrison, The sling in medieval Europe. Bull. Primit. Technol. **31**, 74–79 (2006)

8. The Editors of Encyclopaedia Britannica. "sling", *Encyclopedia Britannica*, 29 May. 2013. https://www.britannica.com/technology/sling. Accessed 25 May 2025
9. U. Singh, *A History of Ancient and Early Medieval India: From the Stone Age to the 12th Century* (Pearson Education India, Delhi, 2008), p.272
10. S. Cuomo, The sinews of war: ancient catapults. Science **303**, 771–772 (2004)
11. J.G. Landels, *Engineering in the Ancient World* (Constable, London, 1978)
12. M.J. Schiefsky, Technē and method in ancient artillery construction: the Belopoeica of Philo of Byzantium, in *The Frontiers of Ancient Science: Essays in Honor of Heinrich von Staden*, ed. by B. Holmes, K. Fischer (De Gruyter, Berlin, 2015). https://doi.org/10.1515/9783110336337
13. D.B. Campbell, *Greek and Roman Artillery 399BC–AD 363* (Osprey, Oxford, 2003)
14. D.B. Campbell, Ancient catapults: some hypotheses reexamined. Hesperia **80**, 677–700 (2011)
15. W. Soedel, V. Foley, Ancient catapults. Sci. Am. **240**, 150–161 (1979)
16. E. Schramm, *Die antiken Geschutze der Saalburg, 1918* (Reprint, Bad Homburg, 1980)
17. A. Wilkins, *Roman Imperial Artillery: Outranging the Enemies of the Empire* (Archaeopress, Oxford, 2024)
18. H. Belloc, *The Modern Traveller* (1898). https://www.gutenberg.org/files/61521/61521-h/61521-h.htm
19. A. Wilkins, Scorpio and Cheiroballistra. JRMES **11**, 77–101 (2000)
20. R. Payne-Gallwey, *The Crossbow* (Holland Press, London, 1903)
21. E.W. Marsden, *Greek and Roman Artillery* Historical Development. (Oxford University Press, Oxford, 1969)
22. E.W. Marsden, *Greek and Roman Artillery* Technical Treatises. (Oxford University Press, Oxford, 1971)
23. C.W.C. Oman, *The Art of War in the Middle Ages* (Cornell University Press, Ithaca, NY, 1953)
24. M.S. Fulton, *Artillery in the Age of the Crusades* (Brill, Leiden, Netherlands, 2018)
25. M.S. Fulton, *Artillery in and Around the Latin East*, Cardiff University Ph.D. thesis (2016). https://orca.cardiff.ac.uk/id/eprint/87056/1/2016fultonmsphd.pdf
26. M. Denny, Optimum onager: the classical mechanics of a classical siege engine. Phys. Teach. **47**, 574–578 (2009)
27. M. McDonough, "trebuchet", *Encyclopedia Britannica*, 21 Feb. 2025. https://www.britannica.com/technology/trebuchet. Accessed 25 May 2025
28. S. Hjelsvold, S. McCallum, *Exarc.net* (2018). https://exarc.net/ark:/88735/10360
29. L. White, *Medieval Technology and Social Change* (Oxford University Press, Oxford, 1962)
30. J. Needham, China's Trebuchets, Manned and Counterweighted, in *On Pre-Modern Technology and Science: Studies in Honour of Lynn White Jnr.* ed. by B.S. Hall, D.C. West (Undena Publications, Los Angeles, 1976), pp.107–45

31. B. Liu, et al., The dependence of baseball lift and drag on spin. J. Sports Eng. Tech. **239**, 1–7 (2022). https://doi.org/10.1177/17543371221113914
32. P.E. Chevedden et al., “The trebuchet”. Sci. Am. 66–71 (1995)
33. A.Y. al-Hassan, D.R. Hill, *Islamic Technology: An Illustrated History* (Cambridge University Press, Cambridge, 1986)

8

They Don't Make 'em Like They Used To

Chapter Summary Two groups of people who build reproductions of ancient projectile weapons, or use them for competition or hunting, are described. The experimental archaeologists seek to better understand ancient technology; hobbyists make or use ancient weapons for competition or hunting. Representatives of these groups, and their views on their subject, are presented. The 'quiet eye' notion of enhanced hand-eye coordination, and a possible evolutionary influence on humans by ancient weapons of long standing, is speculated upon.

"Once a new technology rolls over you, if you're not part of the steamroller, you're part of the road."—(Stewart Brand, American writer.) [1]

1 Living History

I might have appended to the title of this chapter "but a goodly number are trying to do so." There are two broad groups (which overlap somewhat) of people today who are interested in making and using in some manner the ancient projectile weapons that we have investigated in this book. These people—along of course with their constructions—will form the subject matter of this chapter (Fig. 1).

The larger group is made up of (minimal to myriad) hobbyists who are interested, for the most part, in acquiring the skills necessary to throw or fire their weapon of choice. They interact with each other online and face-to-face at fairs and competitions. Some are interested in the history of their weapon but most are interested in the weapon itself, what it can do and how they can learn to use it most effectively.

M. Denny, *Slings and Arrows*,
https://doi.org/10.1007/978-3-032-08563-4_8

Fig. 1 So photogenic. Thanks again to Marion Amalric of Château de Castelnaud, France for this image of counterweight siege engines

The smaller group is made up of experimental archaeologists—amateur or professional—who want to reproduce the weapons of old (these are the "goodly number"). Some go further, and want to make them as they were made in days gone by. Their aim is not necessarily to be especially proficient at making or using their weapon, but rather in understanding how it came to be, historically, and how it performed. In this way they can check the claims made by historical sources about, say, the range of a ballista throwing a one-talent stone ball, and can gain insight into historical battles and sieges where their weapon was employed. Some within this group are craftsmen who want to perpetuate traditional ways of making a weapon—I am thinking here in particular of bowmakers in several countries where bows have great cultural importance.

Both groups include interesting and knowledgeable folk who bring a broad spectrum of skills to the discussion, skills that are not usually prevalent among professional historians or archaeologists—carpentry, hand-eye coordination, engineering, project managing, metalworking, physical strength, stone working, weaving, warfighting.

If you are not a member of one of these groups then you may be very surprised to learn of their size and number. I hinted in the Introduction about this matter by mentioning, for instance, a German society for boomerang throwers. I did not mention that there is a U.K. company that exports between 50,000 and 100,000 boomerangs to Australia, each year. These numbers are likely to be a very small fraction of the total number of boomerangs made because many people like to make their own. Boomerangs fascinate people worldwide today just as much as they fascinated Dubliners in the 1830s. And yet, as we will see, this interest is almost entirely in the returning type of boomerang—the toy, not the weapon of Chap. 3.

Atlatls and slings also fascinate small numbers of people in parts of the world where they were important historically. Thus it might be expected to find atlatl societies in North America or sling aficionados in the Balearic Islands. And archers everywhere—bowmanship and traditional bowmaking are practiced and valued (almost revered) across Eurasia and the New World. Japanese take pride in their unique longbow just as much as Englishmen in theirs, or Native Americans in their short bows.

History similarly casts long shadows for our other weapons: crossbows are still being made and crossbow bolts are still being fired, in Europe and North America especially, though not in China.[1] Bows are used by hobbyists and hunters by the millions today, and this popularity is reflected in the fact that recurve bows, of all our weapons, are the only representatives at the Olympic Games. Siege engines...well, we will get to them; these large, loud, and highly visible poster children of mechanical weaponry appeal to modern people well beyond their historical borders.

It would be madness for me to attempt a comprehensive survey of all the groups of enthusiasts—of both the types that I have described—covering all our weapons across the world. Instead I will highlight a few individuals, societies or events that seem representative and instructive. Let's take them in order.

2 Atlatlists

Professor John Whittaker of Grinnell College, Iowa, is one of the world's experts—and enthusiasts—on everything atlatl. He sent me a 674-page annotated bibliography of research papers involving atlatls from several academic disciplines. He has written many papers on the subject and is an experienced atlatlist. Thus his expertise extents beyond the academic. He keeps score at WAA (World Atlatl Association) competitions. "*World record distance is 848.56 feet... But that was all with specialized modern gear. With normal length atlatls and darts approximating or replicating ethnographic or archaeological gear, a strong thrower can get 100 meters or a bit more.*"[2] Whittaker confirms the statements made in Chap. 1 about atlatl accuracy and lethality: "*Kill megafauna?—Absolutely! I have been doing penetration studies with Pettigrew and team, using carcasses and stone point darts.*" His tests show that atlatl darts can kill animals as large as bison.

I am comfortable drawing conclusions about the physics and engineering of ancient projectile weapons, but when functional factors creep in, or appear

[1] It is illegal for a Chinese civilian to use a crossbow, as they are (and always were) considered to be military weapons.

[2] Readers please note that I am not the only writer on ancient weapons who mixes units—it just goes with the territory.

to do so, then I turn to such experts. The bannerstones attached to the middle of North American atlatls puzzled me, and in Chap. 1 I went along with the consensus view that these weights have some magical or symbolical significance for the thrower. Whittaker agrees with this view, adding the detail that bannerstones are confined to the Eastern and Midwestern regions of North America, and that they provide a slight function for practical atlatlists of balancing the "cocked" weapon. That is, an atlatl loaded with dart and held aloft horizontally by the thrower prior to throwing, feels more balanced in his/her hand with a bannerstone attached.

As for atlatl competitions, in which the general public are encouraged to participate along with seasoned experts, these events are sufficiently mature and established as to have standardized metrics (the WAA's ISAC or International Standard Accuracy Contest). "*Last year I recorded 1393 scores from 81 events in US and Europe.*" Yet despite these impressive numbers Whittaker laments the modest number of hobbyists who participate (perhaps 300) or who consider atlatl throwing to be a hobby (perhaps 1,000). He urged me to come along to an event.[3]

One very interesting nugget of practical information concerning atlatl dart-throwing that emerged from email conversations with Whittaker was this: humans are the best animals at throwing projectiles hard. We are able to aim better at natural targets, perhaps even moving ones, than at paper roundels or bulls-eye targets. This must be a physiological/psychological attribute that has evolved in us, because it has no physics basis. "*Yes, I think that's right and agree it should be physiological or psychological. As a practiced target atlatlist, I know that mental focus is critical, and some of the really good throwers I know are very impressive in focus, good form, and consistent execution of it. Eye focus on the target point is part of this. Hunting is a bit different because you are responding to events in flux, improvising if you like. But I think that spur of the moment action does focus you.*"

Bear this "quiet eye" notion in mind—I will return to it shortly.

3 Slingers

Lewis Everett is an experienced slinger with an interest in the history of his subject. His many Archaic Arms videos attest to his slinging skill; it was Everett's data that I used in the analysis of sling internal ballistics, in Chap. 2. His practical knowledge of slinging is just the type of thing that a desk-bound

[3] I would likely accidentally kill a bystander, so no thank you John. But then it would provide more data on dart lethality...

academic cannot know, and adds to our understanding of ancient projectile weapons. Thus, he assures me that it is not difficult to switch from one sling to another (I am thinking here of Fig. 3 in Chap. 2—a drawing of an ancient Balearic slinger with three slings of different lengths.) Again: we might not know from an examination of ancient sources how tiring it would be for slingers to skirmish in front of an enemy infantry formation—would they become exhausted after 15 minutes or an hour? "*Not tiring at all. Good technique provides high power output with very little effort. A typical slinging session could be ... 6 a minute for several hours straight.*"

He uses many lengths of sling and slinging styles, but seems especially fond of the Balearic. I asked how he thinks he would stack up against a Balearic slinger of old. "*I started slinging just as I entered adulthood, so I never had the benefit from practice since childhood (which accounts for a lot), but have practiced fairly religiously to try and make up for it. Compared to say, the Balearic slingers hired by Hannibal during Rome's wars with Carthage, I think I would utterly pale in comparison.*"

Archaic Arms videos show a man-size target at short range (40–50 ft) being hit more often than not (with the cry *diana!*—"bullseye" in Spanish). Longer distance targets seem not to be a part of slinging competitions today, but Everett believes some of the claims for slingers reliably hitting small targets at 60 m.

He has a thoughtful and cautionary view on modern slinging practice as compared with that of the past. There is no culture today with an unbroken and continuous history of using the sling since antiquity, he points out, and so there is no corpus of accrued knowledge, built up over centuries, that could have been passed down to modern slingers. "*As a result, everyone who practices the sling in the modern day is trying to figure out for themselves how to hit a target, let alone with power, and that's completely aside from sling design...*" This reinvention makes slinging different from archery, but is exactly the problem that experimental archaeologists face when trying to reconstruct ballistas or trebuchets from written sources.

There are likely tens of thousands of slingers in the world who count it as a hobby, entering competitions and posting (sometimes very impressive) YouTube videos, and likely many more shepherds around the world who still use the sling to keep predators off their flocks. The Balearic Islands proudly maintain slinging skills through competitions, and there are many other slinger communities around the world. Everett's views on his target slinging may sound familiar: "*Flat target accuracy is difficult, as it does not excite the brain enough to channel all of one's skill, especially when the focus is extended over many throws. This is part of the reason why many historical accounts across the globe attribute incredible accuracy to the sling. In the context of hunting or self defence, the proportional accuracy will be much higher than when shooting at a roundel target.*"

These are Whittaker's words from a slinger's perspective. I had not prompted Everett with any suggestions about the nature of accurate aiming, and I find it almost eerie that he volunteered an expert's opinion with the same hallmarks, albeit worded quite differently: it is easier to hit a natural target that you have deeply focused on. It makes sense: if a potential target is trying to kill you, it gets your full and undivided attention. If you and your family are hungry, then you will scrutinize a deer or rabbit more than a piece of paper with a bulls-eye on it.

Perhaps these early projectile weapons, the ones that humanity has been using for a thousand generations or more—javelins, atlatls, and slings—have influenced our design as much as we have influenced theirs. What I mean by this is that we have been using these weapons long enough in evolutionary terms for them to have influenced our genetic makeup. There will have been natural selection pressure to pass on to the next generation any helpful random changes that arose in our makeup—an improvement in our perception of moving targets, a hand-eye coordination that more fully involves various muscle groups, an instinctive sequence of muscle actions that is too fast to premeditate. (I am well out of my wheelhouse with this speculation, though it appears that there has been some thinking along similar lines by people who are much more qualified to propound on such things [8]).

4 Boomerangers

There is a World Boomerang Championships held every two years. It began in Perth, Australia in 1991, and the champions that it has produced seem to be mostly German, Swiss, and American. In addition to this international event there are numerous local and regional tournaments held around the world, particularly in the U.S., where dozens of tournaments are conducted each summer. These events involve many categories of event—accuracy, endurance, fast catch, trick catch, and distance throws.

But most, if not all, of these competitions and events apply to returning boomerangs, not to the non-returning hunting and fighting weapon that we investigated in Chap. 3. "Remember, you are the target!" is a traditional warning to beginning boomerangers, and clearly applies to RBs and not NRBs. The latter figure in Australian popular culture, but not in many places elsewhere.[4]

[4] *Kylie* is an Australian girl's name, and one of the names given to non-returning boomerangs.

I have seen combined atlatl and rabbitstick competitions advertised in the United States, but the involvement of NRBs in the World Boomerang Championships seems to be confined to the "Aesthetics Event," which is restricted to artistic design and craftsmanship.

Despite the low profile, NRBs are still being made around the world by enthusiasts. Benjamin Scott in California shows how he made one in a day—his account is particularly interesting for me because of his emphasis on tuning to increase distance.[5]

5 Archers

There are more archers today than atlatlists, slingers, boomerangs, arbalists, and siege engineers combined, and likely always have been, at least since the Neolithic Age. Archery has been so important for so many cultures that today, when it is no longer an essential weapon of war or tool for hunting, it is still practised in almost every country. There are ten million bowhunters in the United States alone, though many of these use the modern compound bow rather than the recurve bow of interest to us. There are 17.6 million registered recreational archers in the U.S., perhaps half a million in Japan, 350,000 in Germany (where bowhunting is not permitted), 70,000 in France, 40,000 in Great Britain, and 17,000 in Italy. Archery is known to be very popular in other countries for which the numbers are unknown, such as India, China, and South Korea.

Archery is an Olympic sport, where it is restricted to the traditional recurve bow, not the odd-looking modern compound bows beloved by many hunters.[6] For four decades archery did not appear in the Olympic Games due to differences in bows and archery practice from country to country. These differences have been ironed out in the sense of standardized equipment during Olympic competitions, and since 1972 archery has been at the Games, and today Olympic archers compete for five gold medals. The most successful country? South Korea.

Recurve bows in competitions may be adorned with accessories such as stabilizers and sights, which were not part of historical archery. Modern competition recurve bows without these accessories are known as *barebows*. The four

[5] See *"All In A Day's Work" Traditional Boomerang Making* at https://www.youtube.com/watch?v=zGcg9m6Eabg. See also an Australian choosing the right piece of wood for making an NRB: *MAKE A HUNTING BOOMERANG-Ep. #7* at https://www.youtube.com/watch?v=YEhGGm-1TVo.

[6] The most interesting feature of compound bows is the *let-off*: draw weight actually decreases with increased draw distance, reaching a minimum at the full draw length, in marked contrast to historical bows where draw weight increases with draw distance. This feature improves aim and reduces fatigue.

main types of bows used in various archery competitions around the world are barebows, recurve bows, longbows, and compound bows. This variety reflects both the popularity of the sport and the historical variations of bows found in different parts of the world.

"*May the eagle dancing of mighty wrestlers fill the earth, the galloping of fast horses shake the vast plains, and the swift arrows of archer's whistle in the sky. I wish my fellow Mongolians, wise and fateful, a happy Naadam.*" Thus spake President of Mongolia Khurelsukh Ukhnaa at the opening ceremony of the 2024 Nadaam Festival. This celebration of Mongolian culture takes place over three days in July each year; the three sports that are considered most important are wrestling, horse racing, and archery—the latter two unsurprisingly, given the country's history.

The biggest archery competition in the world is the Las Vegas Shoot, an indoor event with nearly 4,500 participants in 2024 at the 58th such event. They came from 57 countries and competed as either professionals or amateurs for $550,000 in cash prizes and scholarships. (Professionals? Scholarships? Yes, archery is still going strong.) The International Kyudo Takai is a tournament celebrating the Japanese longbow tradition; the British Long-Bow Society was founded in 1951 for the specific purpose of perpetuating the traditions associated with the recreational longbow.

And why have so many of the Olympic archery gold-medalists been from South Korea? Of course because Korea has a long association with archery, originally military but also for sport. For at least a thousand years the bow was the most important weapon historically for repelling foreign invaders. The Korean bow (a reflexed composite, made with water buffalo horn) achieved a standardized design centuries ago, and its traditional manufacture by craftsmen is considered an important aspect of Korean culture. The bow is called *gakgung*, which means "bow made with animal horn." A rather beautiful video of a modern bowmaker constructing a traditional gakgung displays the thought that has gone into the design, and the skill that goes into the making, of one of these weapons.[7]

6 Arbalists

The International Crossbow Shooting Union (International ArmbrustschüUnion, IAU) is an international sports organization devoted to crossbow shooting. Established in 1956, it is headquartered in Colombier, Switzerland [2].

[7] See *'Gakgung' production process* at https://www.youtube.com/watch?v=WjgtsGKdhMs&t=22s.

The National Crossbow Association in the United States is based in Lenexa, Kansas, and is devoted to promoting the safe use of crossbows in competitions and for hunting [3]. These two organizations encapsulate the modern use of crossbows: competitions and hunting.

Crossbow competitions are organized by various shooting and archery organizations, with two main disciplines: Match Crossbow and Field Crossbow. Match Crossbow is similar to target shooting with rifles, while Field Crossbow involves shooting at various distances. The International Crossbow Shooting Union (IAU) organizes World, Continental, and International Championships in these disciplines.

Crossbow hunting is permitted in many regions for various game species during designated seasons, often aligning with firearms or special archery seasons. Crossbows are generally treated like firearms or bows in terms of regulations, requiring hunters to be certified and follow specific rules about hunting hours, safety equipment, and transport. In the U.S., regulations vary by state.

European folklore includes expert crossbowmen as well as expert archers. In the English-speaking world we all were brought up on the legend of Robin Hood (the archer) while in central Europe there was William Tell (the arbalist),[8] who supposedly was obliged by a tyrant to shoot an apple off his son's head with one shot from his crossbow, Fig. 2. After he had successfully done so, the Austrian tyrant asked why Tell had withdrawn two bolts from his quiver. Tell replied that, had he killed his son with the first, he would have killed the tyrant with the second.

The Swiss association with crossbows extends to the present day. Andi, a Swiss craftsman, makes a European-style crossbow using traditional materials (wood, glue, sinew, and horn for the composite prod—it's an early bow) but with modern tools. The video presentation of the manufacture[9] conveys the intricacy and complexity of the processes, and the fact that modern tools are used forceably reminds us of the skills (and patience) of the medieval crossbow makers.

[8] Both legendary figures were motivated by political causes. If he really existed, Robin Hood lived in late twelfth-century England and concerned himself with wealth redistribution. Tell certainly existed, in the early fourteenth century, and triggered the rebellion that led to the independence of the Old Swiss Confederacy.

[9] *Making a historical crossbow—with the HIGHEST arrow speed?* at https://www.youtube.com/watch?v=K-ogGdXTGkM. See also the video *Medieval crossbow parts and assembly—full process* by JJCh Workshop of crossbow assembly, at https://www.youtube.com/watch?v=WXDmosw3nGQ, which includes a good explanation of the European trigger mechanism.

Fig. 2 William Tell, shooting an apple off his son's head with a crossbow. Woodcut by Daniel Schwegler (ca. 1480–ca. 1546), Hans Rudolf Manuel Deutsch (1525–1571). Public domain

7 Artillerists

In 1991 the *Wall Street Journal* reported on the performance of a large trebuchet that had been reconstructed from medieval drawings by Hew Kennedy, an eccentric English landowner. Kennedy's 620-acre estate in Shropshire provided a safe proving ground for hurling an odd assortment of missiles: a grand piano (to a distance of 125 yards), an upright piano (151 yards), a 50-kg iron weight (230 yards), dead pigs (175 yards), a dead horse (100 yards). The range achieved launching a small car (a 1975 two-door Hillman) was not reported [4]. A smaller prototype engine a few years earlier had thrown gasoline-filled toilets, leading to reports in local newspapers of "those magnificent men and their flaming latrines."

Amusing enough, and the first of many such constructions for entertainment purposes, posted on social media—interested readers will be able to find half a dozen of these on YouTube, launching rocks, washing machines, etc. More serious reconstructions of trebuchets followed, and today there are working models of traction, fixed-CW and swinging-CW trebuchets in Britain, continental Europe, and in some American university engineering departments.

The reasons for making such machines vary, from entertainment and advertising to education and research. Many of these catapults are not reconstructions of ancient designs, but modern machines made with modern materials and tools. Apart from being fun, such machines can be educational—the physics of their operation is instructive, and a confirmation of the predicted projectile speeds and ranges is useful in High School and undergraduate physics and engineering courses.

But reconstructing ancient machines is in a different league from the building of catapults for amusement or instruction. Design is dictated by ancient sources such as Vitruvius, not by modern understanding of the laws of physics. It is possible of course for a modern engine, say a CW trebuchet, to be better than any historical engine of the same size due to improved materials, and even improved designs (interested readers might look into the "floating arm" trebuchet, one of the bells and whistles I alluded to in Chap. 7 and an entirely modern concept.) Making better machines misses the point, of course: the purpose of reconstruction is to learn how it was done in the past; to learn of the capabilities of ancient machines, not of modern ones.

We met Erwin Schramm and Ralph Payne-Gallwey in Chap. 6—they both reproduced ballistas in the early twentieth century and were the first to do so seriously—with the aim of learning about history—and successfully. After them came a number of reconstructions, mainly in Europe, of many types of ancient catapults. Reconstructions of large torsion engines were and still are uncommon, due to expense and effort. Smaller one-man dart throwers abound, in museums and historical reenactment groups, but big ballistas are rare. In Chap. 6 we saw one that had been made (rather badly) by the BBC; in Fig. 3 we see another example, made by a private carpentry company. We have seen that one of the historical reasons why the onager replaced the ballista was the ease of construction. What was true in the fifth century is true today, and so we can expect that nowadays there are more onager and traction trebuchet reproductions than ballistas. The traction trebuchet of Fig. 4 was built by Siri Hjelsvold for research (mostly—it looks like entertainment and education were also likely involved). This reconstruction, carried out under the auspices of a Danish museum, was achieved using period tools and practices [5, 6]. Practical lessons emerged from Hjelsvold's work that educate historians about the traction trebuchets of old. Thus, it took about a day to build engines of the type shown in Fig. 4, from locally-cut wood. The green beams worked well, but dry beams were prone to breaking under the pullers load. A historian might suppose that these trebuchets were built on the site of a siege in order to minimize transport and logistical outlay, but this experimental reconstruction suggests that a preference for green wood must also have been a factor.

Fig. 3 A large ballista reconstruction. Thanks to Carpenter Oak for permission to reproduce this image. Check out their website *Weapons of mass reconstruction*

Similar practical insights result from all such reconstructions. Historians do not necessarily want to know how far a CW trebuchet will launch a flaming toilet or a washing machine, but learning how far it can hurl a dead pig might be useful, given that early examples of biological warfare involved besiegers throwing rotting carcasses over the walls of a fortification so as to spread disease among its defenders.[10]

[10] The earliest example is perhaps the most notorious. The Mongol army of the Golden Horde besieging the Crimean city of Caffa (modern-day Feodosia), a Genoese colony, was ravaged by bubonic plague. They catapulted corpses of their own soldiers who had died from it over the walls and spread the disease throughout the city. Fleeing Genoese traders reached the Sicilian port of Messina, thus spreading this horrible and (in those days) deadly disease to Europe [7].

Fig. 4 A small working traction trebuchet, about to launch a projectile. Note the flexed beam. Photo credit Siri Hjelsvold 2017

8 Summary

So we see that, yes, some people do still make "em" like they used to. Their motives vary: it's fun, it teaches us about different aspects of our history, it demonstrates physical principles. The history part subdivides: reconstructing ancient weapons tells us how the ancients did it, about their technology and skills; it better informs us about past events (battles, sieges) that involved these weapons; it indicates the reliability of classical sources—so if they make impossible claims about weapons then maybe we have to be wary of their other, historical claims.

References

1. S. Brand, *BrainyQuote*. https://www.brainyquote.com/quotes/stewart_brand_172275. Accessed 7 Oct. 2025
2. https://www.iau-crossbow.org. Accessed 7 Dec. 2025
3. https://www.nationalcrossbowassociation.org/

4. G. Mapes, A scud it's not, but the trebuchet hurls a mean piano: giant medieval war machine is wowing British farmers and scaring the sheep. *WSJ* Tuesday July 30 (1991) A1 and A10
5. S. Hjelsvolt, S. McCallum, Traction trebuchet. *EXARC* (2018). https://exarc.net/ark:/88735/10360
6. T.S. Tarver, The traction trebuchet: a reconstruction of an early medieval siege engine. Technol. Cult. **36**, 136–167 (1995)
7. M. Wheelis, Biological warfare in the 1346 siege of Caffa. Emerg. Infect. Dis. **8** 971–975 (2002). https://doi.org/10.3201/eid0809.010536
8. N.T. Roach et al., Elastic energy storage in the shoulder and the evolution of high-speed throwing in Homo. Nature **498**, 483–487 (2013)

Conclusion

Where do our ancient projectile weapons (more accurately, classes of weapons, and somewhat arbitrarily chosen) fit into the fabric of history? When/how have they been at their best? Which has been the most important to the story of our species, and to individual human beings? Battles have been won and lost due to superior or inferior weapons, and meat has been put on the table or escaped being eaten for the same reason. Here are my short answers—if you disagree with them, I would like to know why. (Really. I would be interested to hear any closely argued contrary opinions.)

1 Atlatls

The best of the bunch for hunting megafauna at close range (say up to 30 yards). An atlatl dart packs a heavier punch than a spear or arrow, and is accurate enough at short distances. Not much use in organized warfare, the atlatl was superseded in cultures that progressed technologically beyond the Stone Ages. Yet it was used very widely for thousands of years and so probably played a significant role in the survival of our species. Along with the sling, it may even have changed our DNA.

2 Slings

Probably best used *en masse* during classical antiquity for skirmishing in front of enemy heavy infantry formations. Rivaled the bow in several times and places

M. Denny, *Slings and Arrows*,
https://doi.org/10.1007/978-3-032-08563-4

(Iron Age Britain, Mesoamerica) and an inexpensive weapon in widespread use over millennia. Comparable to the bow in terms of rate of fire and projectile launch speed (and therefore range). Required a lot of training to use effectively.

3 NRBs

Another ancient weapon of long standing, superseded in post Stone-Age cultures by weapons further down our list. The non-returning boomerang was useful for hunting mid-sized game in open country at short to medium ranges (say up to 100 yards). A niche weapon that disappeared early on in most parts of the world, and thus limited in its historical impact, I included it in this book because (time for your nerd author to fess up) it represents the most advanced practical physics adopted by human beings for millennia.

4 Bows

One of the longest lasting and certainly the most important of ancient projectile weapons (at least since the history of our species has been recorded, and maybe longer). Bows have been perhaps the most significant hunting weapon for tens of thousands of years, and a crucial munition for at least five millennia. Bows have shaped warfare and history in the Old World, principally by being the main weapon of steppe nomads and other horse cultures.

5 Crossbows

Important enough in ancient China to dominate warfare, the crossbow later became a niche weapon, there and in Europe. Its influence on history has faded, though it remained for two millennia a significant hunting weapon.

6 Siege Engines

Originally simply enlarged spear-throwers or bows, siege engines morphed into the gravity-powered behemoths of the Middle Ages. They were wholly superseded by gunpowder weapons, and more quickly than were atlatls, boomerangs or bows. (Slings were arguably the only weapon considered here that was not replaced by firearms, for the compelling reason that they had in most parts of the world already lost out to the bow.) The traction trebuchet lasted millennia, which makes it the most historically influential mechanical siege engine. The counterweight trebuchet was a superbly designed artillery piece that changed architecture (in two senses). Hugely important in their day, but their day was short.

7 Development Network

As to where our ancient projectile weapons weave into the fabric of history, consider Fig. 1. This is a technological equivalent of an evolutionary tree, but you can see at a glance that it is not an evolutionary tree. Evolution happens in the here and now—a mutation causes a species to change and so the tree grows a branch. Branches split but they never join together again, in evolution—this is why we call it a tree. I name the technological equivalent in Fig. 1 a "development network," and it is arranged to indicate how our weapons evolved one from another, how they score in terms of importance to human history, and how much they tip the scale of technological sophistication.

It all starts with Paleolithic sticks and string. There are other materials (sinew in lieu of string, some metal parts, glue, horn, bone) but wood and hemp (substitute your string material of choice) are the main players. Three of our earlier weapons need nothing else: the atlatl and boomerang were all wood (originally, and excepting that ivory artifact found in Poland—other materials were employed later) and the sling was all hemp (or nettle or rawhide or...). The staff sling is a sling-on-a-stick, and from it directly developed the traction trebuchet, then the fixed-CW trebuchet and then the swinging-counterweight trebuchet—this is a major branch of our network.

The onager is a simple combination of the two basic components, and something of an evolutionary (developmental) dead end, as you can see. The self-bow was another simple combination, and the progenitor of the second main branch of our network, leading to composite bows, crossbows, and ballistas.

The arrows suggest lineage. Maybe I should insert a few lawyerly weasel-words at this point to evade the wrath of readers—there's a lot to disagree with in Fig. 1. To take one example, I suggest that the bow evolved directly from a stick and a piece of string. In reality the bow was developed independently many times by different groups of people around the world. Let us say that it arose ten times. Perhaps one time it evolved from an atlatl and a piece of string. Maybe twice it arose from hunters who had a flexible stick in one hand, a piece of cord in the other, and a burning desire to find an easier way to project a dart than the atlatl. Maybe twice it evolved from somebody watching another person with a whip. Once from a basket weaver who was contemplating the pliability of the stems she used to make baskets. Three times the bow evolved

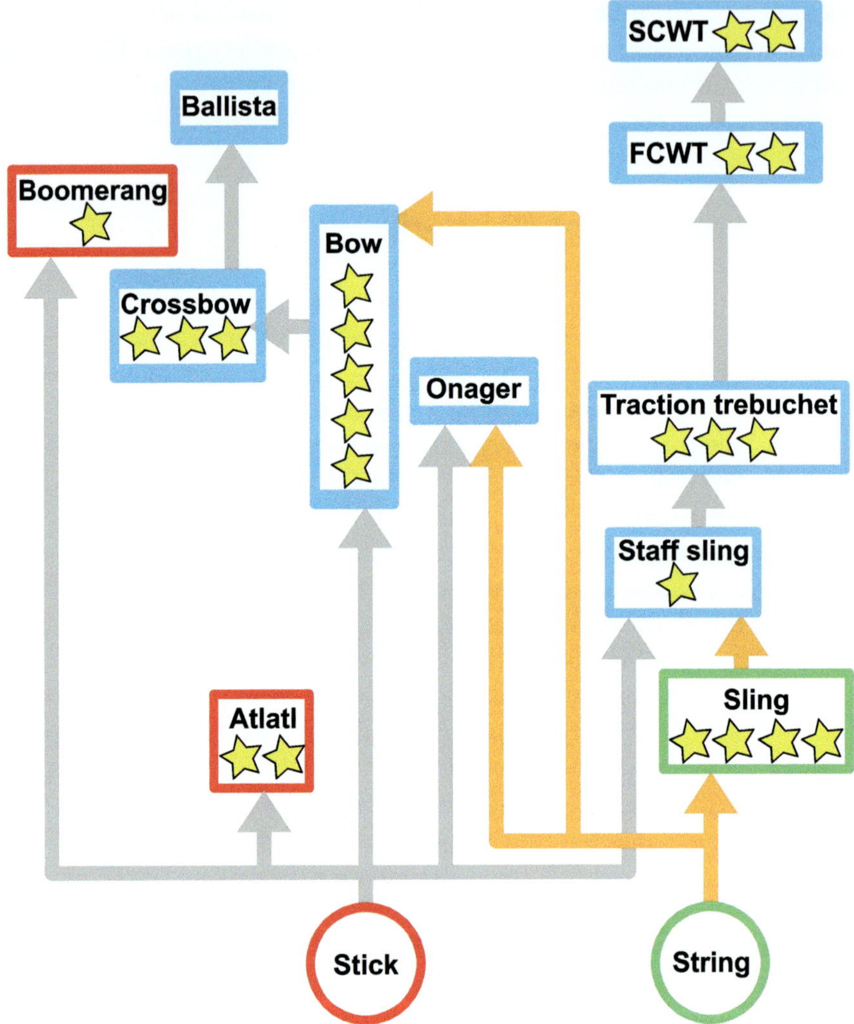

Fig. 1 Ancient projectile weapons development network. The more scientifically sophisticated a weapon is, the higher up the page it appears. Stars indicate (my estimate of) historical importance. A box that is thicker at the top and bottom contains an elastic-potential-energy weapon; a box thickened on the left and right sides contains a weapon that employs gravitational potential energy

from the bow drill (a Paleolithic tool for starting fires). You get the picture: the arrows of Fig. 1 *suggest* lineage, but not necessarily directly or uniquely.

The higher up the page you find a weapon of Fig. 1, the more technically advanced it is—in the sense of construction and, particularly, scientific attainment. We have seen that the science was entirely empirical back in the day, but is no less scientific for that. I rate the boomerang higher (by this attainment metric) than the onager or traction trebuchet despite its simple basic construction—it's a stick. But the physics that we need to understand the flight of a NRB is tricky, and the fine-tuning that is needed to straighten the NRB trajectory is just as tricky, but practically rather than mathematically. The swinging-CW trebuchet is at the top, because of the amount of physics that needs to be mastered in order to make or analyze one. The bow is stretched, vertically, because of the gulf in sophistication between the simplest self-bow and an asymmetric composite bow. Bows went through a lot of development over millennia.

Stars indicate importance to human history. The bow is five-star, topping the list (well, my list) due to its importance in warfare for most of recorded history. The more sophisticated counterweight siege engines are less important because they weren't around for long—all the weapons of this book were eclipsed by firearms, and counterweight siege engines were invented only a few centuries prior to the rise of firearms. The atlatl is important because it figured prominently in many prehistoric societies, for (arguably) longer than any other weapon. This importance did not extent into the historical era.

Technical Appendices

We referred in each chapter to our technical calculations and to their results. Here you will find either a complete derivation (though condensed, as forewarned in the Introduction) or a summary of the key equations with references to my journal papers on the subject. The emphasis both here and (less so, usually) in my papers has been to provide insight via solutions of approximate equations of motion rather than providing more nearly exact answers by brute-force numerical integrations. In part this has been for pedagogical purposes but also because, for many of our weapons (slings leap to mind) it is impossible to write down exact equations due to the complexity of the internal ballistics motion. This complexity arises from the awkward fact that the human body is not rigid and its sources of power (muscles) are distributed, rather than conveniently centralized at one point.

1 TA-0: Aerodynamic Drag and Lift

All our projectiles fly through the air and so are subjected to aerodynamic drag; thus it is worthwhile showing you at the outset its effects upon the range of our darts, sling-bullets, arrows, ballista bolts, and trebuchet balls. Some projectiles are also subjected to aerodynamic lift forces, so I will include these here. The function of the projectile weapons that are analyzed in the text is to generate a velocity for the projectile—a speed in a particular direction. The maximum range occurs when the velocity direction at launch is about 45° to

M. Denny, *Slings and Arrows*,
https://doi.org/10.1007/978-3-032-08563-4

the horizontal.[1] There are three forces that act on a projectile once it is free of its launcher and so is in the external ballistics phase of its trajectory. Gravity, of course, pulls it downward. Drag holds it back—acts opposite to the projectile velocity direction. Aerodynamic lift acts upward, usually, in a direction that is perpendicular to the drag force direction. Of our projectiles only the dart and arrow generate lift forces, and the siege engine ball if it is given backspin (as for a golf ball). Certainly the extreme range of the Ottoman Turks' flight-bow arrows owe much to lift effects. However lift is not a major factor for our darts and arrows. Thus, the fletching on arrows for hunting bows or war bows serves to stabilize the arrow flight, not to generate lift. However it is worth inclusion here because it may be significant for high-speed siege engine balls.

So, we have only three forces to deal with. Even so, the equations of motion that apply for a ballistic trajectory are not solvable analytically—they require numerical solutions. These equations are

$$\ddot{x} = (-b_d \cos\theta - b_l \sin\theta)v^2, \quad \ddot{y} = (-b_d \sin\theta + b_l \cos\theta)v^2.$$

Here x and y are the horizontal and vertical distances traveled, and θ is the velocity direction. We employ Newton's dot notation for time derivative. The full gory details can be found in standard fluid dynamics textbooks [1], and the simpler quadratic approximation which applies for our missiles is also well documented [2, 3]. Here we display the effects on range in the graphs of Fig. 1. The gravitational force has magnitude mg where m is projectile mass and $g = 9.81\ \text{ms}^{-2}$ is the acceleration due to gravity. The aerodynamic drag force is $F_d = -mb_d v^2$ where b_d is a drag factor that is assumed constant for a projectile of mass m, v is projectile speed, and where the minus sign indicates that F_d acts opposite to the direction of v. The lift force is $F_l = -mb_l v^2$ where b_l is the lift factor. Here the constant drag and lift factors are given by $b_{d,l} = \frac{1}{2m} c_{d,l} \rho_{air} A$, where ρ_{air} is air density, $c_{d,l}$ are the aerodynamic drag and lift coefficients, which depend on projectile shape, and A is the cross-sectional area of the projectile, which depends on its mass.

In Fig. 1a, we see the actual trajectory for two choices of parameters: the effects of drag and lift are not significant for low projectile speeds, but drag especially is significant at higher speeds, as here. Note how much the maximum range is reduced. The larger drag factor is appropriate for a spherical 1-lb stone

[1] Less at high speeds, for which the maximum range occurs for lower initial elevation angles due to drag and lift.

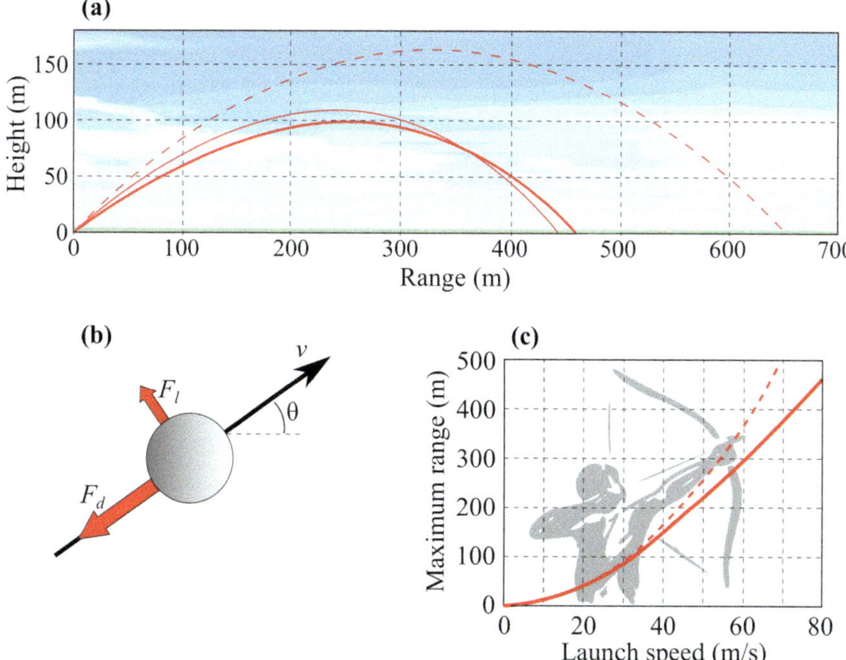

Fig. 1 The effects of aerodynamic drag on projectile range. **a** For launch speed $v_1 = 80$ ms^{-1} and no drag or lift, the maximum range is $R_{max} = 650$ m, with launch angle $\theta_0 = 45°$. With a drag factor of $b_d = 0.001$ m^{-1} $R_{max} = 440$ m and occurs for $\theta_0 \approx 40°$; with both lift and drag (and $b_l = b_d/3$) we finds $R_{max} = 460$ m. **b** direction of lift and drag forces, relative to the velocity direction. **c** R_{max} *versus* v_1 for $b_d = 0.001$ m^{-1} (solid line) and for $b_d = 0$ (dashed line)

projectile, and the smaller one for a 15-kg stone projectile. In Fig. 1b, we see how the drag and lift forces are oriented with respect to projectile velocity. In Fig. 1c we see how range is modified when drag is present. Given that (spherical) projectile mass is $m = \rho\frac{4}{3}\pi r^3$, where ρ is projectile density, and $A = \pi r^2$ we can eliminate projectile radius r. This leads, for example, to the useful formula for a stone projectile $b_d = 0.002m^{-1/3}$, which we will use a lot. Here I have assumed a density for stone of $\rho = 2500$ kgm^{-3} and a drag coefficient $c_d = 0.5$. Similar expressions can be obtained for lead projectiles and for lift factors b_l.

The projectiles of Fig. 1 are stone spheres, for which we have reasonably assumed a drag coefficient of $c_d = 0.5$ and a lift coefficient that is a third of this number. With these values the determined reader can estimate the drag and lift factors $b_{d,l}$ for their projectile mass of choice and from the equations for $\ddot{x}$ and $\ddot{y}$ find the maximum range attainable.

2 TA-1: Atlatl Ballistics

Here you can find the math that led to the number given in Chap. 1 for atlatl capabilities. Some derivations are given; in other cases I simply quote the equations that are derived in my atlatl technical paper [4]. If you need the derivation, you must consult the paper. From the equations given here you can, if interested, reconstruct all the results presented in the text. (This will be the pattern for the analyses of later chapters: simple derivations given in the chapter's Technical Appendix, with more demanding derivations referenced and the resulting equations quoted.)

Let us say that the thrower accelerates a dart a distance d along a straight line (see Fig. 3 of Chap. 1). We can reasonably assume a constant force F throughout the throw, which means that the mechanical energy expended by the thrower is $E = Fd$. If this is all transferred to the dart, then $E = \frac{1}{2}mv_{atlatl}^2$ where m is dart mass and v_{atlatl} is dart launch speed. Equating these two expressions for energy, we obtain launch speed in terms of the force exerted by the thrower. If the distance over which the thrower accelerates the dart is extended by an atlatl of length l, then the launch speed is increased, compared to that without the atlatl, by the factor given in Eq. 1 of Chap. 1.

How big is this force F? We can get an indication from baseball. An MLB fastball pitcher can launch a 0.145 kg baseball at speed of up to 100 mph or 45 m/s. If $d_{pitcher} = 1$ m (eyeballing pitcher action, this seems to be about right—maybe a little longer for some pitchers) and so the force exerted is $F_{pitcher} \approx 150$ N. Most people can't throw a baseball as fast as Justin Martinez can, but a fit person can get to $v_{thrown} \approx 30$ m/s which means they are expending about half as much energy during the pitch as the MLB player. So, we will assume that an atlatl thrower can manage 75 N force which means he could throw a 0.454 kg rock (that's 1 lb) at $v_{thrown} \approx 18$ m/s. This is the number I used in Eq. 1 of Chap. 1.

3 TA-2: Sling Equations of Motion

To calculate the launch speed of a sling bullet we need to model the actions of a slinger during the internal ballistic phase when she is accelerating the bullet. This is complicated because many of the sling styles are complicated, but we can restrict ourselves here to the two-hand pirouette style which is simpler. Here the sling motion can be modeled as circular motion in a fixed plane. Say that the slinger accelerates the bullet for n rotations at a constant angular rate $\ddot{\theta} = \alpha$, starting at $t = 0$ so that, at time t after pirouetting begins, the bullet has angular speed $\dot{\theta} = \alpha t$. It is released after traversing an angle $\theta_n = 2n\pi = \alpha t_n^2/2$. Thus the acceleration lasts for a time $t_n = 2\sqrt{n\pi/\alpha}$. The bullet launch speed is $v = \dot{\theta}_n r$, where r is the radius of the circle (equal to arm length plus sling length). Thus $v = 2r\sqrt{n\pi\alpha}$.

The average power that is transferred to the bullet by the slinger during this wind-up phase is given by $\bar{P}_n = E/t_n$ where E is the bullet energy. It is straightforward to show that the gravitational energy is two orders of magnitude less than the rotational energy, and so here I will neglect the influence of gravity on sling internal ballistics. (Of course gravity matters when discussing external ballistics, but not internal ballistics.) Thus $\bar{P}_n \approx mv^2/2t_n = (n\pi)^{1/2} mr^2\alpha^{3/2}$. We now eliminate α and express the launch speed in terms of average power transferred to the bullet by the slinger, yielding Eq. 1 of Chap. 2.

It turns out that the same result applies if we change the assumption about constant angular acceleration. This interesting fact is derived in [5] and its significance is that it means Eq. 1 of Chap. 2 applies for a wide range of sling styles, and not just for the two-handed pirouette motion analyzed here.

Equation (2) of Chap. 2 can be derived from Fig. 6 of that chapter. The force F that acts is a centripetal force of magnitude $F = mv^2/r$ and persists for a short time δt while the bullet is "unwrapping" from the sling pouch. The torque acting on the bullet is thus $G = Fc = I\dot{\omega}$ where c is the prolate-spheroidal bullet minor radius and $I = \frac{1}{5}m(c^2 + d^2)$ is its moment of inertia about its long axis. Assuming that the pouch covers half the bullet circumference then $\frac{1}{2}\dot{\omega}(\delta t)^2 = \pi$. Putting all this together, and specifying $d = 2c$ we find that the spin frequency $f = \dot{\omega}\delta t/2\pi$ is given by Eq. (2). We also find that $\delta t = 1/f$ and so is a few milliseconds.

The Miller twist rule is

$$S_{Miller} = \frac{30\,m}{T^2 L\left(1 + \frac{L^2}{D^2}\right)},$$

where T is the optimum rifle barrel twist rate in inches per turn, L is bullet length in inches, D is bullet diameter in inches, and m is bullet mass in grains. Converting to sensible metric units this becomes

$$S_{Miller} = \frac{3.8\,m}{d\left(1 + \frac{d^2}{c^2}\right)} \frac{f^2}{v^2},$$

in the notation of Chap. 2. Here I have substituted $T = 39.37v/f$. For an oblate spheroidal bullet with $d = 2c$, and with f given by the spin frequency determined here, the equation for gyroscopic spin stability factor S_{Miller} simplifies to Eq. (3) in Chap. 2.

4 TA-3: NRBs

Boomerangs are very simple machines (only one moving part) but it is the nature of aerodynamics that the equations governing the motion of boomerangs are complicated—much more so than those of more complex machines such as trebuchets. The calculations here are based on those I published in a pedagogical physics journal, [6] using the great simplification of the actuator disk model. (Again, half of physics research is making the right approximations.)

First we will derive an equation for the radius of an RB trajectory, as a test of the actuator disk model. In Fig. 2 we see an RB from the thrower's viewpoint (thick line) initially at some small angle ϕ. So the RB velocity is directed into the page. The lift force of magnitude F_L points perpendicular to both the velocity and the plane of rotation. For level flight we require $F_L \sin\phi = mg$ and $F_L \cos\phi = \frac{mv^2}{R}$ where v is RB speed and R is the radius of its (circular) trajectory. For the actuator disk model we calculate that the torque acting on the RB about its center is $G = -\frac{\pi}{4}\rho c_L v\omega l^4$ where ρ is air density, c_L is aerodynamic lift coefficient, ω is RB angular rotation rate, and l is the length of each RB arm. Torque can also be written as $G = -\frac{v}{R} I\omega\cos\phi$ where I is the RB moment of inertia about its center. Putting all these equations together we

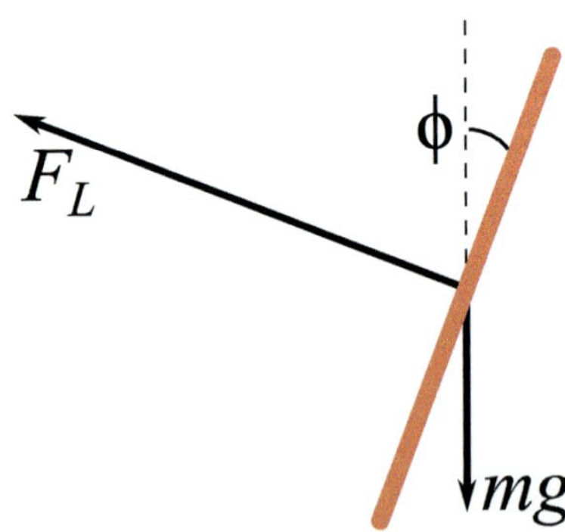

Fig. 2 Forces acting on a returning boomerang (thick line, viewed edge-on, from behind). F_L is the lift force, mg is the gravitational force. Initially ϕ is small, as shown

obtain for the trajectory radius $R = \frac{4}{\pi}\frac{I}{\rho c_L l^4}\cos\phi$.[2] Note that the RB orbital radius is predicted to be independent of ω and v, and is only weakly dependent upon its launch angle ϕ so long as ϕ is small. These predictions are borne out by observation (See Hunt's webpage), and so we take the actuator disk model to be reasonably representative of reality. So now we can go on to apply it to NRBs.

The equations of motion for an idealized NRB (for which the only torque G_D acting is the drag force slowing down rotation) are: $m\ddot{y} = F_D$, $I\dot{\omega} = G_D$, $m\ddot{z} = F_L - mg$. Here we are working with a coordinate system (**x**,**y**,**z**) where **z** points up, and **y** is the velocity direction. Calculating actuator disk model torque, lift force and drag force [6] substituting and integrating we obtain

$$y(t) \approx \frac{2}{k_D}\ln(1 + \frac{1}{2}k_D V t),$$
$$z(t) \approx H + \frac{k_L}{k_D}V\left[t - \frac{2}{k_D V}\ln(1 + \frac{1}{2}k_D V t)\right] - \frac{1}{2}gt^2.$$

For the parameter values of Chap. 3, the trajectory resulting from these equations is plotted in Fig. 4 of Chap. 3. (Note that the constants $k_{D,L}$ are related to lift and drag coefficients $c_{D,L}$ via $k_{D,L} \equiv \frac{\rho c_{D,L} A}{m}$ where A is actuator disk area.)

Now for the concise mathematical explanation of precession (refer to the box ***Precession*** of Chap. 3 and the associated figure). The frisbee without spin acquires angular momentum $\mathbf{j} = \mathbf{G}dt$, i.e., the change in angular velocity over a small time interval dt is just the applied torque multiplied by the time interval. Thus the generated angular momentum is in the direction of $\mathbf{G}$. The frisbee with initial spin $\mathbf{J}$ develops, after a short interval dt, angular momentum $\mathbf{J} + \mathbf{G}dt$; the magnitude J of $\mathbf{J}$ has barely changed but the direction has altered, tilting toward you. Thus the plane of the frisbee tilts backward and the frisbee climbs. This is precession—a slow change in the direction of angular momentum, resulting in a response to applied force that is perpendicular to the direction moved if there were no initial angular momentum (rising *versus* banking left).

In the text I referred to a method of mathematically tuning an NRB, to see how tuning worked. This can be achieved by positing an aerodynamic lift component that varies with position along the leading edge of each NRB arm, i.e., $c_L(r)$ where r varies from zero (the boomerang elbow) to l, the arm length. For such a lift coefficient, we can calculate the torque that acts on the NRB due to lift (see [6]) and then set it equal to zero. This procedure yields the

---

[2] See Prof. Hugh Hunt's Cambridge University engineering webpage *Boomerang Theory* at https://www3.eng.cam.ac.uk/~hemh1/boomerang_theory.pdf.

necessary r-dependent lift coefficient for a tuned NRB, one that does not veer from a straight trajectory due to lift torque:

$$c_L(r) = 6c_L\left(1 - \frac{5}{4}\frac{r}{l}\right).$$

This equation shows that lift coefficient should be negative at the outer edges of the NRB arms, perhaps surprisingly. It would be interesting to see if real NRBs exhibit this characteristic—as yet there is no data to support or refute this prediction.

5 TA-4: Bow Forces and Bow Efficiency

In Fig. 6 of Chap. 4 we posited a very simple bow for which the force required to draw the bowstring was a linear function of the draw distance. Thus the force necessary to brace the bows of Fig. 6 is $F_{brace} = k(b + b_{1,2})$. The draw weight (the force required to pull the string back to the full draw length x_0) is $F_{draw} = k(x_0 + b_{1,2})$. In between, when the draw distance is x, the force is $F = k(x + b_{1,2})$. We can also write this force as $m\ddot{x} = m\dot{x}\frac{d\dot{x}}{dx}$. Assuming that no energy is lost during the draw and release, this same force acts to accelerate the arrow, so that by equating these two expressions for force we find the equation of arrow motion:

$$\dot{x}\frac{d\dot{x}}{dx} = k(x + b_{1,2}).$$

Integrating this equation yields the arrow speeds quoted in Eq. (1) of Chap. 4.

The equation of motion for the spring bow of Fig. 8 of chap. 4 is derived in my bow paper [7]:

$$(mx'^2 + 2I)\ddot{\phi} + mx'x''\dot{\phi}^2 + 2k\phi = 0.$$

Here I is the moment of inertia of each bow limb, ϕ is the limb deflection angle, $x = l\sin\phi + L\cos\theta$ (where l is limb length and L is half the bowstring length) is draw distance, $x' = \frac{dx}{d\theta}$, etc. For this model I assume perfectly stiff limbs, zero bowstring mass, arrow mass $m = 25$ g, and a torque acting on each spring of $G = lP\cos(\theta - \phi) = k\phi$, where P is the string tension. This equation of motion cannot be solved analytically; numerical solutions are shown in Fig. 4.8c,d for a draw weight of 60 lb. If the bowstring mass is m_s then it turns out that bow efficiency ε is reduced to

$$\varepsilon = \frac{1}{1 + \frac{1}{3}\frac{m_s}{m}}.$$

This is about 91% for a dacron string of mass 7 g, but will be significantly lower for a hemp or sinew string.

6 TA-5: Han Crossbows, Mostly

First, I will provide a rough estimation of how effective range depends upon projectile momentum. This was an important factor for arbalists but also for archers, both of whom needed to know the maximum range that their weapon could penetrate armor. Simple assumption number 1: armor penetration is determined by projectile momentum. (Obviously it depends on other factors such as duh, armor. Also projectile head characteristics and projectile energy, but here momentum is taken to be the principal factor.) Specifically I will assume here that if the momentum p of a bolt (or arrow) exceeds some minimum value p_{mail} at range R_{mail} then the bolt will penetrate mail armor at all shorter ranges, whatever the other bolt parameters happen to be. Similarly there may exist a shorter range R_{plate} below which a bolt with momentum p_{plate} will penetrate plate armor.

The energy of our bolt is initially $E \approx \frac{1}{2}mv^2$, where m is bolt mass. Thus $p = mv \approx \sqrt{2mE}$. The energy is constant for a fixed draw length x_0: $E = \frac{1}{2}kx_0^2$. For this rough estimation, I make assumptions 2 and 3 and will ignore crossbow efficiency ε and aerodynamics drag effects, as these vary only slowly with p. Thus range depends on the initial bolt elevation angle θ via the familiar range equation (without drag): $R \approx \frac{v^2}{g}\sin 2\theta$. Note that $p^2 R \approx \frac{4E^2 \sin 2\theta}{g}$ is constant—independent of momentum. (In reality it varies slowly, due to inefficiency and drag). Thus if a crossbow bolt can penetrate a given type of armor with bolt momentum p from a distance R, then it can penetrate a different type of armor requiring bolt momentum p' from a distance R', where $p^2 R = p'^2 R'$. This result is employed in Chap. 5.

Now for a calculation that is specific to Chinese crossbows. I make the same simplifying assumption as for bows that draw weight is proportional to draw length, but will now include drag and efficiency effects. The energy of a spanned crossbow can be then written $E = \frac{1}{2}Fx_0$, where F is the draw weight. If ε is the crossbow efficiency (so that the bolt energy upon release is $\frac{1}{2}mv^2 = \varepsilon E$) then the bolt launch speed is $v = \sqrt{\frac{F}{m}x_0\varepsilon}$. We saw in Chap. 5 that, for Chinese crossbows of the Han period, bolt weight was a constant fraction of draw weight: $mg \approx F/1900$. Thus $v \approx \sqrt{1900\varepsilon x_0 g}$ and the maximum range of such a crossbow is $R_{max} = \frac{v^2}{g}\eta \approx 1900\varepsilon\eta x_0$. Here η is the factor representing range loss due to aerodynamic drag, so that $R_{max} \approx \eta R_0$ with R_0 the maximum range assuming no drag. Thus we expect Han crossbow range to be proportional to draw length, and is relatively insensitive to other parameters.

It is difficult to carry out a detailed analysis of the Chinese multi-prod crossbows, so here is a rough overarching estimation of optimum number of prods, for a multi-prod crossbow consisting of n identical prods, as a function of crossbow efficiency. Say that a single crossbow stores energy E_1 when spanned, and transfers energy εE_1 to the bolt, where ε is efficiency. We expect the efficiency ε_n of an n-prod bow to be about $\varepsilon_n \approx \varepsilon^n$ and so the bolt energy will be $E_n \approx n\varepsilon^n E_1$. The multi-prod bolt gains energy, compared with the single-prod bow, by a factor $G_n \equiv \frac{E_n}{\varepsilon E_1} = n\varepsilon^{n-1}$. For realistic choices for single-prod efficiency, this calculation leads to the results shown in the table.

Multi-prod energy multiplication factor G_n

ε	0.5	0.55	0.6	0.65	0.7	0.75	0.8
1	1.00	1.00	1.00	1.00	1.00	1.00	1.00
2	1.00	1.10	1.20	1.30	1.40	1.50	1.60
3	0.75	0.91	1.08	1.27	1.47	1.69	1.92
4	0.50	0.67	0.86	1.10	1.37	1.69	2.05
5	0.31	0.46	0.65	0.89	1.20	1.58	2.05

Thus, for low crossbow efficiency $\varepsilon \leq 0.5$ arbalists should stick to single-prod bows; for $0.55 \leq \varepsilon \leq 0.65$ a two-prod bow is more efficient; and for $0.7 \leq \varepsilon \leq 0.75$ a 3-prod bow is the best choice. A 4-prod bow is the most efficient when each of the component prods is of the highest realistic efficiency.

7 TA-6: Ballistas

We need to calculate the launch speed of a ballista projectile. In Fig. 3 we have (half) an idealized ballista with stiff frame and arms as shown, and with spring constant K. Constraints on the variables θ, ϕ due to the geometry are $d + l\cos\phi = L\sin\theta$ and $Y = Y_0 - l\sin\phi - L\cos\theta$, where $Y_0 = l\sin\phi_0 + L\cos\theta_0$. Initial angles are ϕ_0, θ_0 and final angles are $\phi_1, \pi/2$.

The energy of the system during the internal ballistics phase is given by $E = \frac{1}{2}mv^2 + I\dot{\phi}^2 + K(\phi + \alpha)^2$ where α is the pre-stress angle (see Chap. 6 text). Assuming that energy is approximately conserved, so that initial stored

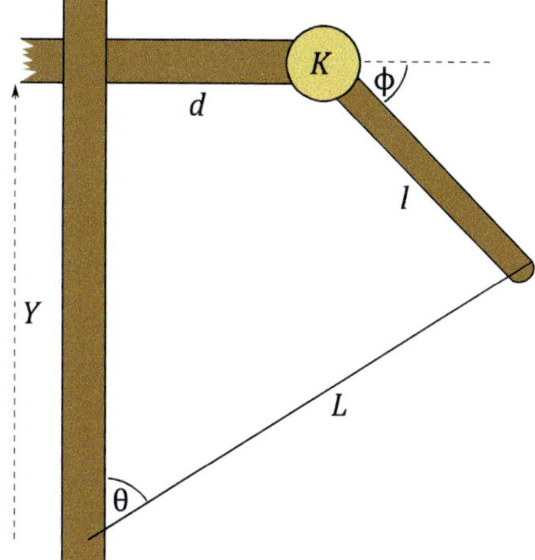

Fig. 3 Half an idealized ballista, with dimensions labeled. *K* is the spring constant

energy equals energy of the system at release $E_0 \approx E_1$, we obtain for projectile release speed

$$v_1 \approx \sqrt{\frac{2K}{m}(\phi_0 - \phi_1)(\phi_0 + \phi_1 + 2\alpha)}.$$

Here we have assumed that initial angular speeds are $\dot{\theta}_0 = 0, \dot{\phi}_0 = 0$. If $\phi_1 \approx 0$ then this equation reduces to the expression shown in Chap. 6.

From the foregoing it is straightforward to calculate the fraction of initial stored energy is taken away by the projectile at launch—in other words, the ballista efficiency. This is

$$\varepsilon = \frac{(\phi_0 - \phi_1)(\phi_0 + \phi_1 + 2\alpha)}{(\phi_0 + \alpha)^2}.$$

Equation (12) tells us that efficiency falls as ϕ_1 increases, and so it should be as close to zero as possible. In this case ε reduces to Eq. (2) of Chap. 6.

8 TA-7: Staff Slings, Onagers, and Trebuchets

There's a lot of calculations packed tightly into this section. Let's begin with the staff sling. In Fig. 4 we have an initially horizontal staff (length a) that rotates clockwise about its lower end at a constant angular speed ω. At first the sling (length b) dangles vertically; thus the initial staff angle is $\theta_0 = \pi/2$ and

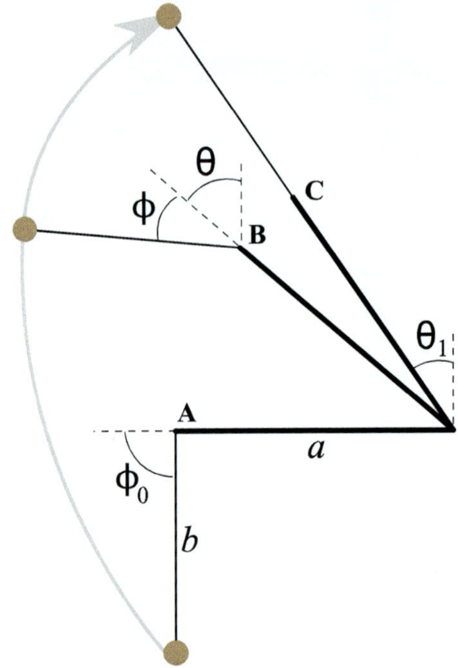

Fig. 4 For our staff-sling calculation: staff (thick line) and sling (thin line). The initial position is at A and the projectile is launched at C when staff and sling form a straight line

the initial sling angle is $\phi_0 = \pi/2$. The projectile is released at $\phi_1 = 0$ when $\theta = \theta_1$. In between, the angles are θ, ϕ. The staff motion progresses from A to B to C. Measured from the pivot point the projectile position is (X, Y) where $X = -a\sin\theta - b\sin(\theta+\phi)$, $Y = a\cos\theta + b\cos(\theta+\phi)$. The sling equation of motion is $\ddot{\phi} = -\frac{a}{b}\omega^2\sin\phi - \frac{g}{b}\cos(\omega t - \phi)$, where g is, as usual, the acceleration due to gravity. Here I have used the fact that staff angular speed is constant, so that $\theta = \frac{\pi}{2} - \omega t$.

For typical staff-sling parameters we expect $a\omega^2 >> g$ and so we can neglect the gravitational term in the equation of motion, which thus simplifies to $\ddot{\phi} \approx -\Omega^2 \sin\phi$. But this is just the familiar (to physicists) equation for a simple pendulum with angular frequency $\Omega = \sqrt{\frac{a}{b}}\omega$. For small angular excursions the solution is familiar to every high school student, but here the angular excursion ϕ is not small, so we employ the less-well-known but impressively accurate *harmonic balance* approximation [8, 9], which tells us that the angular speed of the sling is $\dot{\phi} \approx -\frac{\pi}{2}\omega' \sin\omega' t$ where $\omega' = 0.850\frac{a}{b}\omega$. From Fig. 4 we see that the projectile release speed is thus $v_1 \approx a\omega[1 + r(r + 1.335)]$ where $r^2 = \frac{b}{a}$ is the ratio of sling to staff lengths. Had we launched the projectile from a staff of length $a + b$ with no sling, its launch speed would have been $v_0 = \omega(a + b)$ and so, finally, we obtain the ratio v_1/v_0 of staff-sling launch speed to "lever" launch speed:

$$\frac{v_1}{v_0} \approx \frac{1 + 1.335r + r^2}{1 + r^2}.$$

This is the equation that gives us the numbers quoted for "double-pendulum effect gain" in Chap. 7.

For completeness I should show how sling length is fixed by the desired release angle. It is easy to show from the foregoing calculations that the projectile is launched at angle $\theta_1 = \frac{\pi}{2}(1 - 1.176r)$ and so if, for example, we want maximum range ($\theta_1 = \pi/4$) then $r = 0.42$ whereas for maximum launch speed ($\theta_1 \approx 0$) then $r \approx 0.85$.

The early onager—without a sling—is easy to analyze, with the usual assumptions about rigid arms and negligible internal friction [10]. The potential energy stored in the spring, when the arm is pulled back, is $E_0 = \frac{1}{2}K\theta_0^2$ where K is the spring constant. The energy after release of the trigger is $E = \frac{1}{2}K\theta^2 + \frac{1}{2}I\dot{\theta}^2 + (\frac{1}{2}M + m)gR\cos\theta$ when the arm is at angle θ (see Fig. 5 for notation). Here $I \approx \frac{1}{3}MR^2$ is arm moment of inertia about the pivot. Energy is approximately conserved during the short interval of arm motion; solving $E = E_0$ yields the projectile launch speed v_1 given in Chap. 7

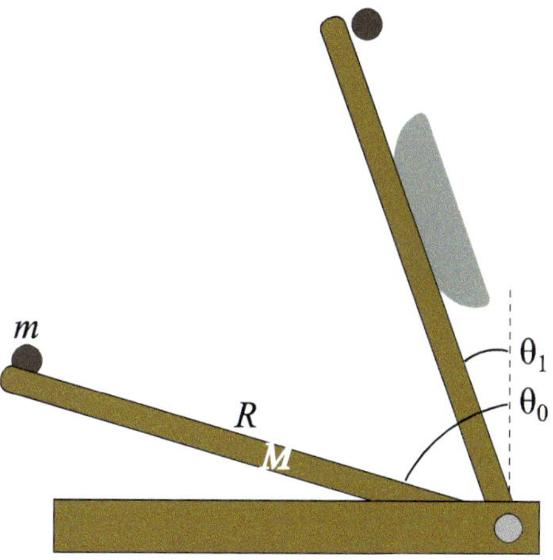

Fig. 5 Onager, at the extremes of its arm motion. θ_0 is the braced (fully drawn) arm angle, and θ_1 is the angle of the buffer. R and M are arm length and mass; m is projectile mass

(neglecting gravitational energy—it is easy to show that this is negligible compared with spring potential energy and projectile kinetic energy).

From here on, the calculations are condensed bare-bones outlines requiring advanced classical mechanics. The traction trebuchet shown in Fig. 6 has beam length L, and mass M, an axle located a distance fL from the rake end, a force (arrow) directed downward, a projectile located at (X, Y) measured from the axle, and a projectile mass m. Initially the beam is horizontal but then moves (to angles θ, ϕ as shown) under the action of the force. Projectile location is $X = -a\cos\theta - r\sin\phi$, $Y = a\sin\theta - r\cos\phi$ where $a = (1 - f)L$. Projectile speed is $v = \sqrt{\dot{X}^2 + \dot{Y}^2}$. Trebuchet kinetic energy is $T = \frac{1}{2}mv^2 + \frac{1}{2}I\dot{\theta}^2$. (The sling of length r is considered massless.) To obtain the projectile equations of motion we employ the generalized form of the Euler-Lagrange equation:

$$\frac{d}{dt}\left(\frac{\partial T}{\partial \dot{q}}\right) = \frac{\partial T}{\partial q} + F_q, \quad q = \theta, \phi.$$

Here F_q is the generalized force associated with the variable q. Our q are angles and so the generalized forces $F_{\theta,\phi}$ are torques about the beam axis and sling rotation axis, respectively. After some calculation we find for the equations of motion [11]:

$$\begin{aligned}(I_b + ma^2)\ddot{\theta} &= mar\ddot{\phi}\sin(\theta - \phi) - mar\dot{\phi}^2\cos(\theta - \phi) + F_\theta, \\ r\ddot{\phi} &= a\ddot{\theta}\sin(\theta - \phi) + a\dot{\theta}^2\cos(\theta - \phi) - g\sin\phi.\end{aligned}$$

Here I have substituted $F_\phi = mgr\sin\phi$, and I_b is the moment of inertia of the beam about its axle. F_θ consists of the puller term $-fLF_n\cos\theta$ (plus minor contributions from the beam and projectile) where F_n is the downward force applied by n pullers. Numerically integrating this pair of differential equations yields the solutions presented in Chap. 7.

For the fixed-counterweight trebuchet, the equations of motion are the same but with a different F_θ: the downward force of the pullers is replaced by the counterweight $M_{cw}g$. In Chap. 7, we also adopt different initial conditions than for the traction trebuchet.

Equations describing the swinging-counterweight trebuchet are more complicated because there are three degrees of freedom θ, ϕ, ϕ_{cw} and because there is a constraint (the rail, described in the text and shown in Fig. 6). These equa-

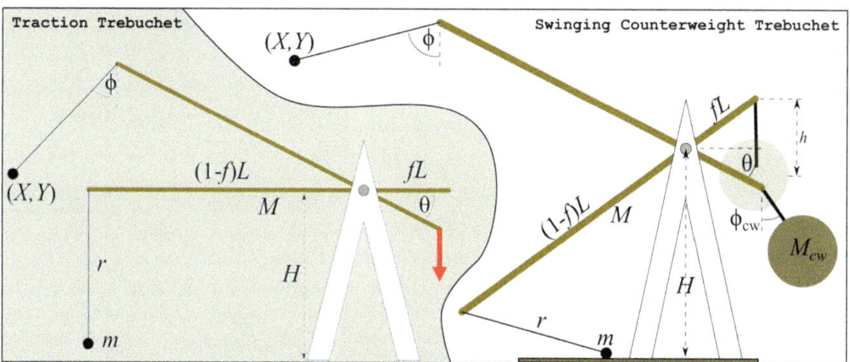

Fig. 6 Left: Traction trebuchet model and notation. The beam and projectile are the only moving parts with mass. The system is shown in its initial stationary configuration $(\theta, \phi) = (0, 0)$ and later during the launch phase. Right: Counterweight trebuchet model and notation. Now there are three moving parts and three variables $(\theta, \phi, \phi_{cw})$. Also, the projectile is constrained to slide along a horizontal rail until it becomes airborne. Again, initial position is shown, and a later position during launch

tions are derived in my paper [12] and this derivation will be omitted here. They are

$$\begin{aligned} I\ddot{\theta} &= M_{cw} f L h \ddot{\phi}_{cw} \sin(\theta + \phi_{cw}) + m(1-f)Lr\ddot{\phi} \sin(\theta - \phi) + V \cos\theta + G(\theta, \phi, \phi_{cw}), \\ h\ddot{\phi}_{cw} &= fL\ddot{\theta} \sin(\theta + \phi_{cw}) + fL\dot{\theta}^2 \cos(\theta + \phi_{cw}) - g \sin\phi_{cw}, \\ r\ddot{\phi} &= (1-f)L\ddot{\theta} \sin(\theta - \phi) + (1-f)L\dot{\theta}^2 \cos(\theta - \phi) - g \sin\phi, \end{aligned}$$

where $G(\theta, \phi, \phi_{cw}) = M_{cw} f L \dot{\phi}_{cw}^2 \cos(\theta + \phi_{cw}) - m(1-f)Lr\dot{\phi}^2 \cos(\theta - \phi)$. Here V is the gravitational potential energy of the system and I is its moment of inertia about the beam axle. These equations describe the engine motion when the projectile is free of the rails; when it is sliding along the rails then there is a fourth equation involving a Lagrange multiplier, for which I refer the interested reader to Ref. [12].

References

1. B.S. Massey, *Mechanics of Fluids* (Chapman and Hall, London, 1989). (ch. 8)
2. E.L. Houghton, N.B. Carruthers, *Aerodynamics for Engineering Students* (Arnold, London, 1982)
3. J.L. Bradshaw, Projectile motion with quadratic drag. Am. J. Phys. **91**, 258–263 (2023)
4. M. Denny, Atlatl internal ballistics. Phys. Teach. **57**, 69–72 (2019)

5. M. Denny, Internal ballistics of the sling. Am. J. Phys. **93** 367–375 (2025). https://doi.org/10.1119/5.0226263
6. M. Denny, Non-returning boomerangs: an engaging exercise in rotational dynamics. Eur. J. Phys. **46**, 025002 (2025). https://doi.org/10.1088/1361-6404/adb65f
7. M. Denny, Bow and catapult internal dynamics. Eur. J. Phys. **24**, 367–378 (2003)
8. M. Denny, Harmonic balance: simple and accurate estimation of nonlinear oscillator parameters. Eur. J. Phys. **42**, 065014 (2021)
9. M. Denny, Watt steam governor stability. Eur. J. Phys. **23**, 339–351 (2002)
10. M. Denny, Optimum Onager: the classical mechanics of a classical siege engine. Phys. Teach. **47**, 574–578 (2009)
11. M. Denny, Siege engine dynamics. Eur. J. Phys. **26**, 561–577 (2005)
12. M. Denny, Siege engine dynamics. Eur. J. Phys. **26**, 561–577 (2005)

Videography

A number of enlightening and entertaining videos pertaining to the subject matter of this book have appeared on YouTube in recent years. I recommend these to you while the subject is still fresh in your mind.

Chapter 1:
– *Atlatl Basics with Matt Graham.*
– *Atlatl Demonstration.*
– *How To Build An Atlatl (Spear-Thrower).*
– *The Atlatl: Most Underrated Stone-Age Tool?*

Chapter 2:
– *Extreme Sling Velocity.*
– *Sling Hit 175 m (574ft).*
– *Slinging Physics—Release Timing and Spin Influences.*
– *Slinging Spin Rate.*

Chapter 3:
– *"All In A Day's Work" Traditional Boomerang Making.*
– *Aussie versus American Hunting Boomerangs.*
– *How to Throw an Aboriginal Boomerang.*
– *Hunting Boomerang | Rabbit Stick.*
– *MAKE A HUNTING BOOMERANG—Ep. #7.*
– *Rabbitstick!*
– *The Rabbitstick: The Not So Primitive Hunting Weapon.*

M. Denny, *Slings and Arrows*,
https://doi.org/10.1007/978-3-032-08563-4

Chapter 4:
– *Ancient Archery Workshop in China That Still Makes Bows and Arrows.*
– *The Archer's Paradox in SLOW MOTION—Smarter Every Day 136.*
– *Flemish Twist Bowstring (Quick and Easy).*
– *"Gakgung" Production Process.*
– *Genghis Khan and the Mongolian Genocide.*
– *How to Make a Bowstring for a recurve bow?*
– *How to Make a Longbow String for Beginners No Jig Required ...*
– *How to Make A Sinew Bowstring for a Primitive Survival Bow. Primitive Technology.*
– *How to Make a Traditional Korean Composite Bow.*
– *Let's Get It Right: Longbow versus Crossbow—A Video Essay.*
– *Positive Tiller, Explained.*
– *Making an Insane Penobscot Bow—Relaxing Bow Build.*
– *Slow motion archery, archer's paradox, arrow spine.*
– *What Was the Most Lethal Weapon Of The Dark Ages? | Samurai Bow | Chronicle.*

Chapter 5:
– *Chu Ko Nu Chinese Repeating Crossbow History.*
– *Crossbows—Spanning Methods*
– *Crossbow Trigger Designs #2.*
– *Gastraphetes.*
– *Making a Historical Crossbow—With the HIGHEST Arrow Speed?*
– *Medieval Crossbow Parts and Assembly—Full Process.*
– *Physics of Ancient Chinese Crossbow*
– *The Goats Foot Lever.*
– *470lbs Belt and Lever Medieval crossbow.*

Chapter 6:
– *Ancient Roman Ballista Tests on Different Settings.*
– *Ballista: A History of the Weapons that Played a Decisive Role in the Expansion of the Roman Empire.*
– *Ballista Development through History.*
– *Can This Ancient Roman Catapult Live Up to its Reputation?*
– *Greek Ballista.*
– *The Manubalista Was Rome's Secret Weapon in Ancient Britain.*

Chapter 7:
– *Unleashing a Medieval Trebuchet on a Wooden Palisade.*
– *How to Build a Trebuchet/MythBusters.*
– *Medieval Trebuchet Hits Wall... and the Wall Wins ?!?.*

– *Trebuchet Siege Artillery—Battle Castle with Dan Snow.*
– *Trebuchet (How Does It Work?) The Largest Siege Machine During Medieval Age.*

Chapter 8:
– *How to Throw an Atlatl | Live Free or Die: DIY?*
– *The Most Underrated Ancient Projectile.*
– *How This Guy Became a World Champion Boomerang Thrower | WIRED?*
– *Traditional Archery Tips—How to Shoot a Recurve Bow?*
– *Historical Crossbow Triggers.*
– *Young Ballista on the Track (BALLISTA BUILD PT.3).*
– *Small Trebuchet Firing.*

Online Resources

Here are three websites that I have found to be particularly useful on Japanese and Chinese military history:

– Luca Bottazzi's library platform is at: https://gunsenmilitaryhistory.wordpress.com/

– *Great Ming Military* at http://greatmingmilitary.blogspot.com

– *A Crossbow Mechanism with Some Unique Features from Shandong, China* at http://www.atarn.org/chinese/bjng_xbow/bjng_xbow.htm

Wikipedia articles are of variable quality, but a number of the history entries are a good place to start; here are two: *Battle of Agincourt, History of crossbows.*

M. Denny, *Slings and Arrows*,
https://doi.org/10.1007/978-3-032-08563-4

Further References

These came to light after I had finished the book or somehow didn't make it into the relevant chapter bibliography. I have not read all of them beyond the abstract and conclusions but all were recommended to me and look to be valid, relevant, and interesting:

T. Dorshorst, "Archery's Lasting Mark: A Biomechanical Analysis of Archery," (2019). Masters Theses. 827. https://doi.org/10.7275/15119161 https://scholarworks.umass.edu/masters_theses_2/827.

H. Knecht, "The History and Development of Projectile Technology Research," in *Projectile Technology*, H. Knecht ed., pp. 3-36. (Plenum, New York, 1997).

K.-H. Hsaio and H.-S. Yan, "Mechanisms in Ancient Chinese Books with Illustrations," (Springer, Cham, Switzerland, 2014).

S. LeBlanc, Steven, *Prehistoric Warfare in the American Southwest*, (University of Utah Press, Salt Lake City, 1999).

C. Lepers, J. Coppe, and V, Rots, "The propulsion phase of spear-throwers and its implications for understanding prehistoric weaponry," *J. Arch. Sci.: Reports* **59** 104768 (2024).

A. Milks, "Skills Shortage: A Critical Evaluation of the Use of Human Participants in Early Spear Experiments," *EXARC* Journal Digest 32-34 (2019). https://exarc.net/ark:/88735/10426.

J. Ortiz *et al.*, "Experimental and Computational Study of Archery Arrows Fletched with Straight Vanes," *Proceedings* **49** 56 (2020). https://doi.org/10.3390/proceedings2020049056.

M. Denny, *Slings and Arrows*,
https://doi.org/10.1007/978-3-032-08563-4

M.B.Vega, N. Craig, "New experimental data on the distance of sling projectiles," *J. Archaeol. Sci* **36** 1264-1268 (2009). https://doi.org/10.1016/j.jas.2009.01.018.

J.C. Whittaker, D.B. Pettigrew and R.J. Grohsmeyer, "Atlatl Dart Velocity: Accurate Measurements and Implications for Paleoindian and Archaic Archaeology," *PaleoAmerica* (2017). DOI: 10.1080/20555563.2017.130113.

L. Zieliński, "New insights into Nubian archery," *Nub. Arch.* (24) 791-801 (2015). https://doi.org/10.5604/01.3001.0010.0164.

Index

M. Denny, *Slings and Arrows*,
https://doi.org/10.1007/978-3-032-08563-4